FACTORS INFLUENCE FUTURE TECHNOLOGICAL MARKET DEVELOPMENT

JOHN LOK

ISBN 979-888629095-0

Contents

Preface

Preface

This book divides three parts. The first part explains what future technology factors will influence US economy development. The second part explains what future technology factors will influence UK economy development. The final part that I shall give some economy theories to explain why these kinds of technological factors can influence future these both countries' economy growth. It brings this question: What are future influential technological development trend to US and UK?

This book concerns to be given on my opinions and some economists' opinions to UK and US both countries government and businessmen what will be the most urgent important needs to develop on the technological product investments aspect. I shall indicate the online teaching technology and environment protect technology, such as natural energy, farming environment protect technology etc. as well as automatic technology in manufacturing industry , such as human intelligence products to give my reasons and some economists' reasons to explain to support why these three technological developments will be valid to invest to develop to UK society in the future.

In this book final part, I shall give reasons to explain why some economy theories can be proved to satisfy these technological factors requirement to predict which can influence US and UK future economic growth. The economy theories include: welfare economic theory, Kaldor -Hicks efficiency theory, Pareto improvement theory, industrial production and pollution economy theory, natural rate of unemployment theory. Finally, I hope readers can attempt to analyze whether these economic theories can be applied to explain whether these technological factors can assist US and UK future economic growth.

How economists measure developing and developed countries by technological influence. It will be valid to be judged whether these three new technological development will be the most urgent important needs among other technology development to UK and US society within ten year. I hope that our ideas can assist UK and US government and businessmen to make the right decision to concentrate on developing these three technological development within ten years.

In my this book chapter one, I shall explain what technology will influence

UK future development? Nowadays, UK is one effort technological development country. I believe it has effort to develop any technology, e.g. building, space navigation, bioscience, computer, medical etc. different technological development. However, in the economic theory view point, it indicates our earth has limited and scare natural resource to supply human to use, so if UK government or businessmen made wrong decision to choose to concentrate on developing any one of the non-important or urgent needs of technology, then UK country will face to use any scare natural resource to develop the important or urgent needs of technology to provide UK citizen and their next generation to satisfy their future needs. Thus, how to allocate to apply the natural resource to develop which kinds of high technology issue which is one valid research question to UK society nowadays.

In fact, UK building technology had reached the mature stage, the building technology innovation is very excellent to compare other countries. So, UK government does not need spend more resource to invest to building technology aspect. Also, the biotechnology technology is excellent to satisfy any hospital patents' needs. Even, computer technology and space navigation technology are also excellent to compare other countries. UK had invented many different kinds of new computers or space navigation products to satisfy UK citizen's needs. So, UK government does not need to encourage to provide financial support to any UK businessmen to develop these two kind technologies. What kind of technologies will be UK government and businessmen who need to spend to invest and develop within ten years. I believe that environment and education and automatic manufacturing technologies will be UK future new technological development trends.

During economic development stage, any country must encounter any new challenges and these new challenges had not encountered to occur to any country in the past. However , the most fast economic growth of country, such as US, it will have possible to encounter these same challenges during its economic development stage.

I write this book another aim to give my view points to indicate and explain what factors will cause US future economic growth.

In this book chapter two, it will explain why these factors will impact US future economic growth. The factors include external environment impact of developing countries cities technological competitive investment factor, the trends impact on rural America's future economy factor, high level education and high birth rate factor , socio-economic and political factor

, an increase in the returns to education factor influences US future labor market change, popular science, technology , engineering and mathematics kind of labor supply will be increased demand in US, increase development in genetics, human intelligence, robotics, nanotechnology, 3D printing and biotechnology technological industry factor, US entrepreneurship innovation influence US future geography economic growth factor, the role of intangible assets influences the regional economic growth in US factor, social factor impacts US future economic growth, tourism industry influences US future economy growth, the effects of population growth influence US economy growth, reducing income inequality factor influences future boosting US economic growth, long term cheap medical cost trend factor influences US economic growth, talent management factor influence US economic growth factor, the impact of educational quality factor, bio-medical industry factor.

Prologue

UK Future Unique Technology

Online teaching technology

Future, online teaching method will be popular to be applied to teach to any university, even secondary and primary schools. Because internet service is free charge to any students in any countries. Many different age students who can know how to apply internet as well as internet studying is very convenient to any students who can to internet to learn or study in home or public library or school library conveniently. Teachers do not need spend much time to teach students in classroom. They can use internet to teach teachers by face to face seeing and talking to their individual student from every student's computer. So, students do not also often spend much time to go to school to learn. So, developing any fast speed and time saving and talking and listening online teaching methods will be popular needs to any UK primary and secondary and university students in the future. It will be one new technological teaching method to change the traditional classroom educational method in UK and schools. For example, when one UK student who had left UK and is living in another country long time. If any UK school did not provide online teaching service to any UK students. It means that the UK citizen can not choose study himself/herself any UK school if who still hope to study any UK course when who is living in another country. Even one foreign student who does not go to UK to study, if he/she can find any UK primary or secondary or university to study from online. Then, the UK school won't lose one foreign student, due to it does not provide online teaching method to any foreign students. So, online Technology educational learning method will be one popular learning method which is enhanced, supported, mediated or assessed by the use of electronic media. Technology also enhanced learning may involve the use of new or established technology and/or the creation of new learning material. It may be deployed both locally and at a distance (i.e. a combination of traditional and e-learning approaches), to learning that is delivered entirely online. Online learning technology characteristics (features) include identification of a project lead for each area of any learning strategy, identification of two " quick win" for example lecture capture, electronic submission and feedback.

How can online technology enhance learning at UK any schools? It will include these several aspects to analyze. On identifying, prioritizing and innovation hand, online technology is a process for resourcing, prioritizing, acquiring and evaluating school software and hardware for UK any school needs. On staff development learning plan and a student skills development plan hand, UK schools need to establish a base-line policy on the standard (minimum) technology enhanced learning expectation for education each program and module and a mechanism for updating the schools' policies. On evaluation and research hand, a mechanism for engaging the owners of the technology enhanced learning strategy with best practice in the sector including contributing to and benefiting from pedagogical research and the evaluation of the student experience to UK any school.

Thus, UK schools can apply online technology to develop on educational aspect, such as digital literacies and appropriate technical skills that equip UK students for life-long learning, graduate level employment and professional practice, be empowered to learn how to learn with online teaching technology, using online technology to engage in interactive, creative and co-constructed learning with the potential for online learning in an interdisciplinary and international context, using online teaching technology to engage in learning with and from people from anywhere in the world, be supported on placement and in workplace learning through mobile applications and other supportive technologies that facilitate their online learning when away from the classroom, having access to innovative methods of online learning teaching and assessment that are the foundation of a research-lead academic environment, engaging with UK schools in developing , implementing and reviewing the technology enhanced learning strategy. Thus, in the future, it is important to build a capacity to the online education strategy to adopt future learning innovation and student individual online learning need (demand) to UK any school online teaching trend.

There are many examples where UK academics working in isolation or in small UK teaching organizations or classroom learning groups have developed teaching innovation that have a positive impact on UK students' academic experience , but these have remained isolated to particular modules or occasionally program. The aim of education researching online learning process is to identify the good online teaching innovation that is being developed and to prioritize those that have the potential to make a significant contribution to improving the academic student experience at UK any schools. This online teaching process will need any UK schools which can plan how to apply limited resources necessary to achieve online teaching. In addition, the online teaching research process would evaluate and prioritize large scale educational software and hardware requests for primary, secondary and university students' requests. An important part to this process will be to ensure the integration of online educational products and packages that school staff and students regular use to make routine working and access as seamless as possible.

Decisions about school administrative online technologies should not be taken in isolation before assessing the impact on UK teaching staff. In addition, a range of techniques such as, online expert facilitation, coaching and peer support will be used to support individuals, groups or longer academic units, who are learning on major technology enhanced online learning projects. Staff engagement may also facilitated through incorporating technology that is used in teaching staff research and/or professional activity that can be cooperated into their teaching.

Online learning technology can develop UK students skills, UK schools need to understand how UK students understand technology and learn with it, therefore the digital literacy strategy needs to be considered as part of the overall strategy as well as the relevant skills development in UK employability strategy. So, in the future, online learning strategy will make it clear that students will develop technical skills the appropriate level for graduate employability and professional practice. Also, in the future, the online technology can enhance learning working group to discuss external development, that are of educational strategic importance, understanding and evaluating current best practice and research and understanding and evaluating the online educational strategic contribution that pedagogical research and student feedback can have on online educational strategy, policy and practice. The e-learning unit is responsible for informing and educating. This could be done by, for example, providing a short digest of relevant information for each meeting and by setting aside a proportion of each school meeting to discuss a topic of particular online educational strategic interest to every school. Academics that have not got a specialist interest in online educational technology enhanced learning will need relevant information at an appropriate time. This could be provided at a school department or faculty level and this will have clear links to the staff online teaching development plan. Hence, online educational development strategy will influence any UK educational school technological improvement in the future.

Environmental protection technology

Why does environment technology valid to UK businessmen and government to develop? Nowadays, global air and water pollution is serious. Even, UK has many farming is polluted by the water and air pollution. It will influence UK farmers' income if whose farm land (natural resource) is polluted by water or air (natural resource). Even it will influence UK citizen will encounter food shortage if UK farmers can not grow any fresh and health food to provide the enough food numbers to eat every day. Moreover, air and water pollution will influence UK citizen drink the polluted water and breathe the dirty air to live every day. This natural resource (air and water challenge) will influence UK citizen health to cause illness , even death every easily. So, UK government can not neglect the natural environment pollution challenge. The environmental protection technology will help the UK and development countries to solve the challenge of climate change to avoid or reduce farming, foods, or vegetable or fruits or rice, pork, livestock numbers loss threats, i.e. the development and deployment of low carbon energy technology, including technology for the efficient use of energy. The commercialization of low carbon energy and energy efficiency technologies in the UK, with a specific focus on the demonstration and deployment phases of bringing low carbon technologies to UK market.

The UK Government needs to deliver a low carbon economy and to meet UK ambitions emission reduction target. So, low carbon and environmental protection technology researching and development will reduce the carbon intensity of energy production as well as reduce energy demand, towards meeting the contributing UK's ambitions production as well as reduce energy demand, and renewable energy goals. The use of energy (including transportation fuel) and the UK's targets on climate change, for example, by helping the UK make a step change in increasing deployment of renewable energy, improving UK energy efficiency and helping low carbon technologies reach the market. The development of low carbon technologies, and to realize the benefits of doing so in terms ensuring security of energy supply for the UK future economy development.

In UK, private sector investment in technology innovation in the low carbon energy sector will other sectors of the economy. So, in UK energy technologies are likely needed to be developed to avoid dangerous climate change, or an acceptable cost. So, in the future, UK government will need to consider to research environment protection and low carbon energy technology. The activities will reduce carbon emissions, or have the potential to reduce carbon emissions on the longer term, through the use of energy technology will accelerate development and deployment of low-carbon energy and energy efficiency technologies will capacity in the demonstration and deployment of low carbon technologies. Innovation in the energy sector is the only way to identify, develop and reduce the costs of new and improved technologies for the extractions, generation, distribution and use of energy. It has long been an important means of achieving the UK's energy policy aims of a secure and affordable energy supply, as well as to develop the environmentally friendly technologies that are required in UK response to climate change, i.e. nuclear, wind or water, sun energy technology, which is future new energy technology is suitable to research to create to apply instead of current electricity energy.

How global warming influences UK agriculture growth. Scientists have also been fighting the use of chlorine in municipal water systems to kill various strands of bacteria. Chlorine reduces by about 80% the number of alimentary tract diseases relative to polluted, unchlorinated water. A relatively new genetically modified agricultural products. They were partly successful in Europe, such as UK (some countries banned genetically modified products) in spite of the fact that neither history nor research supports their case. People began to modify plants as early as the beginning of the agricultural revolution (8000 to 10,000 years ago), when they started seed selection and who have continued ever since. The green revolution of the 1960 year brought about strains of grans and rice more resistant to a variety of local conditions. The effects have been that countries like India, which had suffered from recurrent famines over the millennia, became self-sufficient in food due to the resultant sharp increase in agricultural productivity. It was a real science and technology over the poverty dominating most of human history. But it is precisely the products of science and technology that ecologists are so deathly afraid of. In an interesting study in a quarter (28%) of clinically analyzed cases of obsessive compulsive disorder were cases resulting from the fear of global warming. To destroy the modern, whether industrial or postindustrial, civilization, human have to destroy an important engine of economic growth, that is its energy sources. And this is what eco-warriors try to achieve under the banner of against global warming. Thus, UK government will have responsibility to attempt to research new technology to fight global warming challenge for itself farmer benefits and even global benefits both on the future.

● Future Economist global warming technological protection economic influence opinion

Some future economists indicated reasons to explain why UK government and businessmen needed to consider how to develop natural environment protection technology to avoid global warming challenge to influence UK economy development. They indicated the anthropogenic (human-made) global warming resulting from the increase in "greenhouse gas". They offered their perspectives on the scientific valid of anthropogenic global warming phenomenon, its probability of occur and expected consequences and is dominated by technologists, economists and political scientists, who considered the need to make the horribly costly adjustments in energy generation and usage suggested by climate alarmists.

Many stress that global warming is primarily caused by other phenomena than human use of fossil fuels or human activities in general. They are looking at the activities of the sun and impact of the larger universe as the main source of global warming and stress that global warmings (plural) happen intermittently with global cooling. I shall explain why global climate warming will influence to the political and economics of the issue to UK country. For it is the latter, rather than the global warming itself, that will pose a challenge to the Western world, such as UK and the world at large in the future. Scientists concerned who should move forward with policy measures to avert the alleged disaster. They also apply manufacturing theories to support enough to frighten politicians into action and scare societies into acceptance of measures that would sharply reduce UK citizen their living standards. Otherwise, UK politicians had support that bureaucracies were established, money allocated and lobbies created dependent on the new kind of subsidies. In consequences, climate alarmism and resultant interventions in national economies and human activities have become the increasingly wide spread and increasingly cost reality. With the growing availability of money distributed, and even more promised, a range of benefit of the global warming machinery has been on the increase. So, if UK government did not concern how to innovate new weather protection technology to avoid climate change adverse (poor) influence. It is possible that billions of dollars of UK public money are needed to spend on research global warming challenge because global warming will influence UK agricultural industry. UK agricultural industry is one important export income source to raise UK GDP income every year. If global warming become very serious to influence UK weather to be bad to cause UK farmers who can not grow good taste food and vegetable to supply to domestic and overseas food consumers to eat. Then, UK will loss much GDP income from local agricultural export sale. It seems global warming and agricultural production which has direct relationship to influence UK economy development in the future.

The main problem with climatology is that it must be based as already stressed on very many variables affecting climate and too few hard data necessity. Differences apply not only with respect to the scale of changes obtained, but even to their direction (rising or declining temperature). Some weather scientists indicated to concern global warming challenge. In consequence, it would be impossible to discover if and where errors were made not only in estimating relationships between variables but also in the quality of data used. (Hauser, J. Tellis, G. J; Griffin, A. 2006) They were comparing average temperatures measured some 30, 40 or 50 years ago by, say, 90 % weather stations in the countryside and 10 % stations in the cities with contemporary average temperatures measured by weather stations located today on 50:50 basis in the countryside and cities. Then, one could obtain the increasing temperatures without any real world climate or even weather changes. Comparability would be ensured if the same number of countryside-located and city-located weather stations had been compared for different periods. The alarmists intentionally mix up " temperature growth" with the trend of temperature growth. To give an example, if in the first decade the temperature grew by 0.5 % degree, in second decade it grew by 0.3 % and in third decade it grew by 0.1%, what was registered was a growth in the temperature, but certainly not a trend of growing temperature. A fourth decade should, on the basis of the trend, bring about no change in the temperature.

To conclude, scientists believed that global warming was caused by human's bad behavior more than natural environment influence. So, it is human's responsibility needs to solve this challenge, due to who feel earning profit aim is more important to protect natural environment, e.g. air and water pollution , due to manufacturing process is the main factor. So, UK has responsibility to attempt to research how to solve global warming challenge , such as it has many famous scientists who can devote their scientific skills to cooperate to solve global warming challenge with other countries' scientists. Some weather scientists also hypothesized that human may be at the end of the present warming period. If they are right, it would be bad for humanity, as warmer periods have always been associated with better conditions for economic activity. To sum up, scientists believed that global warming will influence human economic activity to be bad.

● Climate technology of global warming success to UK farmers
Climate alarmists were able to convince a large part of the Western public and a majority of Western politicians of the cause of fighting against the global warming. It supposes itself in an instinctive preference for collectivist solutions in economic and social spheres, with negative to disastrous consequences when scientists are applied in practice, so UK government needs to concern global warming challenge, due to it is possible that it will influence UK natural environment weather to be poor to influence many UK farmers' agricultural and vegetable and fruit and rice wheat etc. food growth successfully. What is the global warming influence to cause disease? For example, ecological alarmists and activists (eco-warriors) never admit they are wrong, they long pursued their fear mongering campaign against chlorine. Their success in branding DDT a dangerous substance had a negative impact on the malaria eradication campaign in poorer parts of the world. Alternatives to be have been far less effective and the result has been the resurgence of malaria cases and the manifold increase in malaria -caused deaths to the largest extent in Africa.

Automation technology in manufacturing industry
Nowadays, UK computer and space explore technology had reached the mature stage. It means that UK government ought not need to continue spend much resource to research these two kind technologies. Otherwise, the automatic manufacturing technology, e.g. human intelligence new product. It has need to develop because human intelligence machines will bring beneficial to satisfy human everyday life need, e.g. hospital patients' activities need, if the patent who can not walk easily, but the human intelligence machine can assist the patient walk to anywhere conveniently.

So, he/she does not need to sit on wheel chair and apply the human intelligence machine man to help him/her to drive on the intelligence automatic driving vehicle to go to anywhere conveniently.

Otherwise, increased automation in low wage countries, e.g. China, Korea, Africa, Hong Kong etc. which have traditionally manufacturing firms, could use automatic technological manufacturing to bring lose cost advantage and potentially lose their ability of achieving rapid economy growth by shifting workers to factory jobs. So, UK government and businessmen needs to consider automation technology development, i.e. 3D printing manufacturing industry will encourage UK companies to move manufacturing process, closer to gain the biggest advantage from this 3D automation technology development.

A growing concern of premature de-industrialization in energy and developing countries could require new models and a need un-skillful the UK workforce. In the future, the best way toward for UK cities will reduce their exposure to automation is to boost their technological dynamic and attract more UK skilled workers. Automation technology progress can give UK manufacturers' employee benefits, such as long term healthy productivity improvement, raising productivity efficiency and product quality, macroeconomic and microeconomic effects of automation technological change, it's change will be beneficial to UK society, i.e. automation active labor market policies, which could help UK job seekers find jobs from training to incentive to support self-employment to create high technological job employment chance in UK society. So, raising science, technology, engineering and math subjects update skills level are needed to UK any universities, which can be increasingly important in UK society, these factors could complicate the ability of UK high automation technology education to adopt to the UK automation manufacturing technological change. A talent mismatch already exists in UK, with many well UK educated workers can find employment in lower-skilled jobs. To combat this, greater coordination will be needed between the education, training and employment sectors in UK society.

Why are high automatic technology product development models needed to research to UK any manufacturers? UK government and manufacturers need to consider how to achieve high technology product development models. According to Hauser et al. (2006) indicated the high technology (high tech.) development process, is influenced by the innovative process, bringing products on exception value which stimulate product market demand. Innovation provides products the specific basis for which world economies compete with each other on the global market. Able to find new solutions, innovations generate significant changes in existing markets, destroy them, or create new marketing (Hauser et al. 2006). So, UK manufacturers need to concern on any manufacturing high technology product development process because which can influence any new products development to manufacture to sell to any overseas or domestic both markets successfully.

What is high tech. product meaning? Mohr et al. (2010) argues that there are two reasons why it is important to clarify and specific high technology : (1) due to the impact of technologies on the economy, attempts are made to classify economic production and incomes ; (2) due to the impact of high tech. on the environment. Standard marketing strategies are being modified and adopted , therefore, it is necessary to know the products to focus on. Why UK manufacturers need to consider high technological product process. Nowadays, high tech. products are complex, advanced, requiring specific technical knowledge, which is technologically not discontinued and being produced at the companies which have twice as many technical personnel and invest twice as many in scientific research and development than other companies. Moreover, these products are time-sensitive as scientists are continuously searching for new approaches for invention of more advanced technologies which make all preceding ones lower-ranking. The most important, nowadays global consumers will adopt the particular technology. It means that global customers may delay adopting new high-tech. products and in order to mitigate the prolonged uncertainty require a high degree of education and information about the product and need post-purchase reassurance.

Anyway, nowadays customer individual needs in high tech. environments are characterized by sudden changes related to unpredictable fashion. Even, consumers concern about how to preserve new product' competitive technological standard is completely incompatible with technological uncertainty. The most important factor is the prevalence rate of any new products development process, which is influenced by slower than of traditional products. In many cases high-tech. automatic product market are being materialized slower than which are expected. The technological uncertainty challenges will exist in development process, such as uncertainty related to the timetable for development of the question whether the new product will be function as promised. In automatic high-tech. industries, the time requires for product development is difficult to predict as , commonly, it takes longer than expected , uncertainty related to unanticipated consequences and uncertainty about the product life cycle related to competition products. In conclusion, these factors will influence new automatic technology product development process unsuccessful, so UK manufacturers will need to concern on any high technological automatic product's manufacturing process.

Before, all over the world presented picture of demonstrate in London on the occasion of the meeting of the G20. Some economists indicated disastrous economy consequences will occur to any one of Western country , such as UK, so if any one of Western country did not consider automatic technology development to itself country. They indicated one example, such as material incentives to produce disappeared throughout Russia and, when Society leadership called off the experiment, the country faced industrial output reduced to 10% of what had been registered in 1914 and agricultural output reduced to such low levels as to cause widespread famine.

Why would UK encounter disastrous economy consequences if UK government did not encourage manufacturers spend money to invest to innovate automatic technology industry? According to a variety of anthropological studies, a collectivity is unable to operate efficiently with everybody giving talent workers have chance to devote whose best effort to manufacture any high technological products, e.g. human intelligence vehicle or airplane. Hence, economic incentives are needed to UK manufacturers to invest high technological automatic industry development. Because the economists predict UK will have many talent worker numbers, their number will be more than a certain number of normal effort workers, due to UK technological education level is very excellent to provide to train many young technological manufacturing students to find this kind of high technological manufacturing job. So, the high technological manufacturing job seekers will increase and it won't decrease to UK job market in the future.

Assuming that UK high technological automatic manufacturing workers who would desire only to introduce changes in the workings of the international economic order and policies of countries participating in the present economic order rather than change the order itself, what will be UK manufacturers their specific economic preferences in the future? It implies tnat either concentrate on spending more investment to automatic high technological development, e.g. human intelligence automatic high technological products or still concentrate on spending more investment to common traditional technological products.

However, UK was a developed Western country which had had strong automatic high technological development effort very long time. Otherwise, it compared to some developing countries, such as Asian China, Hong Kong, Korea etc. Asian countries their future economic growth rate will show un- surprising , different patterns, so the Asian countries has weak effort to invest high automatic technological product development, such as human intelligence technological development. The catching-up process suggests low economic growth rate in the high automatic technological product development to the Asian developing countries in the future.

Hence, the future economists predict that it views as probable successors of the Western world economic leadership if any Western country , such as UK manufacturers who prefer to invest to any high automatic technological products development , e.g. developing on human intelligence automatic technological products more than traditional common technological products development. On the one side, but it seems important to stress that two very poor countries among the challengers-China and India-are examples of countries that changed their institutions and economic policies from no or little economic freedom to more economic freedom. Because there two countries whose governments prefer to lend loans to encourage their country manufacturers prefer to invest high automatic technological products manufacturing. On the other side, attitudes toward foreign direct investment (FDI) have undergone change since the 1960 s and a large majority of less developed countries, e.g. China and India are now competing strongly among themselves and with developed market economies for direct investment from multinational companies. So, UK will face China and India high automatic technological product competitors in the future. And in fact, all countries that joined Western developed economies did that without much (if any) external inflow of public resources. It is right time that UK government needs to lend loans to encourage domestic manufacturers to invest high automatic technological products to raise whose international high technological products sale effort to win its future competitors. So, machine resources will be increased demand to o UK manufacturers if who chose to spend machine resources to innovate to manufacture any new and high technological automatic products to raise human daily life needs in the future. It means that it is right time UK manufacturers need buy much machines to prepare to manufacture many future high technological automatic products when these machine prices are low. Because the future global machine prices will possible be raised if many China and India manufacturers will also buy many machines in the future. For example, USA government had provided much financial support to assist sugar cane producers to develop their businesses. And they are dependent to a much larger extent than sugar cane producers and sugar processors in the USA on government. Without very high subsidies to renewable energy generation, they would not have survived at all. So, USA government had been the first country which could lent much financial assistance to encourage domestic renewable energy generation manufacturers to develop high technological energy manufacturing business. So, UK government needs follow USA to lend financial assistance to encourage domestic high technological automatic industry development. Future economists also predict China and India will be competitors for future leadership in the global economy, special high technological products. China has been the media and analyst's favorite for quite some time. Quantitative projections have seemingly supported such expectation. Such as China and India had manufactured many high technological new space rockets products, ocean war large ships etc. Moreover, China has become one of the major world trade players in the early twenty-first century.

Many long-term forecasts, assuming similarly high economic growth rates in the decades ahead, predict that China will surpass the USA in terms of aggregate GDP somewhere between 2020 and 2030 or later, say between 2030 and 2050 year. The future economists conclude on the basis of these predictions that China will not only pass the USA in aggregate product (GDP), but its economy and economic policies will influence the rest of the world to a similar extent that the USA does at present.

I stressed a very important point, namely that the UK future high technological automatic product competitor China and India, namely that economies not only grow, but in the process change their structure. China and India have been industry very rapidly (the first transition) and building the physical infrastructure that accompanies industrialization changes to technology in the future. However, at a certain per capita GNP level the two countries, such as China and India will face another structural shift when which technological development will reach the mature stage in the future. China and India had been primarily historical pattern of economic development because the shift in the role of engine of growth from industry to services is to a much greater extent a qualitative shift. Both higher and different skills are required. And, even more importantly, interactions generating ideas driving the highly human-capital-intensive service economy require a much freer environment, not only in the economic area. Chinese exports have been heavily labor-intensive. This being the case, they contributed to the expansion of industrial employment, offering for the first time in the history of China a taste of (very modest) prosperity to more than 100 million new industrial workers and their families. This is the major component of the success accomplished by Chinese economic growth. Richer trade partners create room for more trade, so the Chinese should hope that intra-South trade, that is, trade between the emerging economies of Asia, the Middle East, Africa and Latin America, will open up new and growing opportunities. I presume that if Western economy , such as UK did not developed high technological automatic industry to stable their social welfare, so thoroughly slowed down their economic growth.

Will it allow China to accomplish the transition to a mature, innovation, service-sector-based market economy? It has allowed the economy to industrialize much more successfully, even if the labor shift from agriculture to industry has not yet been completed. But it is a long way off the next major test: the second high technological industry transition of the economic structure to China. Bear in mind that Russia attempted it twice and failed at both attempts.

But even, assuming that China at some point in the future does succeed in accomplishing the second transition, will it be able to supersede the USA, for example, as the main global high automatic technological innovation center if it wants to become the No.1 global high technological industry economy? Given the nature of the centralized state and its stability to collect financial resources , China's ability to increase research and development expenditure to high automatic technological products and to hire a mass of researchers, engineers, technicians and other specialists should not be doubted. This process in already taking place.

But , again, Soviet Russia already exceed the USA in the R&D/GDP ratio in the 1970s, long before the communist collapse, with no effects on its innovativeness. Inputs matter less than outputs, quantity in the innovation process mean much less than quality. The latter characteristics depends importantly on economic, civic and even political institutions. Otherwise, independent India had three options open to it in 1946s. It could pursue spontaneous economic development, with some state intervention to be sure, along the lines of basically free market capitalism; it could turn the clock back and try to recreate the rural-agricultural and handicraft based. The dominant way of thinking was Society -style priority to industrialization and , within industralization , priority to heavy industry. In other words, not textiles and clothing, which has been developing well in India since the mid- nine teen century, but production of sewing machines and , even better, production of machines the produce sewing machines.

The results were only to be expected. The heavy stress on the expansion of capital-intensive heavy industries in a very poor country quickly strained the ability of the Indian economy to generate adequate savings. Moreover, some of these industries were above the level of industrial competence of an underdeveloped economy. Thus, the amount of required resources (capital, skilled labor) was usually larger per unit of output than in the same industries in more mature, richer industries economies. In another view point, India will develop light industries, just as any other poor country with a great deal of unskilled labor, had a comparative advantage and no less importantly, an economy in which, due to their low capital/labor ratio, light industries could employ many more people, spreading prosperity more widely in a poor country. So, it explain that why China will have more effort to develop heavy high technological industry in the future. Thus, India got less economic efficiency, less employment than in a spontaneously developing economy, less ability to compete internationally in light industries suitable for an underdeveloped economy and finally got heavy industry unable to compete even on the domestic market and, therefore requiring no less heavy a dose of protection. Overall India got an underperforming economy, in particular in its relations with the rest of the world.

To conclude by comparing the performance of the traditional sectors of the Indian economy and the performance of its modern, human -capital-intensive subsector of manufacturing and skill intensive service sector. The latter both employ workers with high-and medium -high skillful level (in branches ranging from computer software and biotechnology and pharmaceutical high technological light industry). India is ahead of China in terms of the output and export of such products and services. Thus, it implies that UK ought concentrate on developing high automatic heavy high technological industry, e.g. human intelligence technological products because these industry is not better development to other many countries' strong effort , such China and India large population countries.

Reference

Hauser, J. Tellis, G. J; Griffin, A. 2006. Research On Innovation: A Review And Agenda For Marketing Science. 25(6): 687-717.

Mohr, G. J. Griffin, A. 2010. Research On Innovation : A Review And Agenda For Marketing Science. 25 (6): 687-717.

CHAPTER TWO

US Future Unique Technology

Environment protection technology

Some economists predict developing countries every city in the global 750 is projected to have a larger future technological economy growth. But the diversity of developing countries' economic performance is large. Developing economy cities, such as China, Japan, Hong Kong, Korea cities can grow rapidly by acquiring capital and technological know-how and putting them to use by their rapidly growing urban labor forces. Even, these developing countries cities' rapid technological development can impact to US labor market supply.

Due to future rapid technological development to these developing countries, the result will cause developing countries cities, such as Asia China, Hong Kong, India cities economy growth will rapidly. Otherwise, developed western countries cities, such as US, UK, Canada, Australia, Span, Germany etc. countries lie close to the technological frontier have stable urban populations and more limited investment and job creation opportunities. It will influence these developed countries' economic growth is slow than the developing countries. Due to developing Asia countries cities will prefer to invest more technological development to compare to developed countries, such as US, UK, Canada, Australia, Span, Germany etc. countries. Therefore developed countries, such as US, UK, Canada, Australia, Span, Germany etc. countries tend to grow more slowly. It seems developing Asia countries, such as Hong Kong, China, Korea etc. countries urban and central cities economic performance within developing countries will be better than developed countries, such as UK, US urban and central cities within five years.

Thus, I suppose that future developed countries, such as US the speed of technological development will be slower to compare developing countries, such as Hong Kong , China, Korea, India etc. countries. Then, it will bring this question: How to solve developed countries, such as US, Washington, New York etc. cities future economic slow growth challenge? For future developed country cities, such as New York, Washington etc. large cities' technological investment and location decision, it will need understand that diversity is essential. Various factors can have an impact on US country intra-national urban cities performance, including sector structures, agglomeration benefits, infrastructure quality. For example, US country central government needs have tolerance of diverse performance, land supply and US city governance plan. It aims to raise US central and urban cities' future economic performance. Moreover, US country will need to concern above issues to arrange how to improve central and urban city economic development . Specially, it will need to concern technology innovation to encourage many overseas technology investors to anticipate different kinds of technology products innovation, e.g. human intelligence machine, mobile, computer etc. high technology products. It aims to absorb different countries' new technological knowledge of different kinds of high technological products to excite US domestic or foreign countries consumers' desires to choose to buy many different kinds of high technological products to raise US GDP growth in US technological product sector industry in the future.

Agricultural development technology

Some economists indicate that there are five trends reshape to impact rural America' future economy. They include that digital economy will shift future America rural economy. US quality of life will change a lot, the US rural economy will stay uneven, US commodities will compete in global markets and will give less benefit to US rural economy and US new products will revolutionize US agriculture economy.

The first aspect impacts to US agriculture economy, the US future rural economy stays uneven. Growth will concentrate in 4 out of 10 rural places and they have scenery, a retail hub, or one next to a city in US. The impact on rural America includes some rural places will try to manage growth , but many places on a quest for new economic engines. Thus, it will bring these questions to US rural economy impact, such as : Who will be US businessmen clients? The struggling farmers? The struggling farm-dependent country? The booming mountain area? The rural area transforming into city?

The second aspect impacts to US agriculture economy, due to US commodities will compete in global markets if it will bring a smaller benefit in US rural economy. Then, US farm scale will cut costs and competition fewer US farms and places will depend on farm income. What is the impact of commoditization on rural America? There are more farms depend on area jobs and it creates a new imperative to add value. What is the impact of commodity on US clients? The need for competitive commodities remains, but the payoff for added value is rising and community impacts are important.

The third aspect impact to US agriculture economy, new products will revolutionize agriculture. Major, shift from commodities to products, spurred by biotech, means two agriculture in the future and two rural America . Hence, US future agriculture determination the rural economy will be declined. A future new US agriculture supply chain integrator will be caused from the traditional farming supply chain procedure the change to outsourced contractor farming supply chain procedure, such as: In beginning, from farmer will outsource supply chain contract to processor, then contract to distributor and contract to food retailer final step.

The fourth aspect impacts to US agriculture economy, what will be two agricultures impact to the US future economy? The first US agriculture impact will be US commodity agriculture. It focuses on production capabilities, farming foods production will be thin margins maintained with technology and big sale. The second US agriculture impact will be product agriculture. It focuses on consumer needs, farming foods production margins will be protected by capturing value and building business relationship. Thus, it will bring these questions concern what the impact of product agriculture which can influence to US economy. How to apply biotechnological agricultural techniques and farming product application to raise US farming productivity? How to build world class agricultural (farm) producing chains that benefit to US farming produces production method? How can US faming foods producers participation or lead in product chains?

In the fifth aspect impacts to US agricultural economy, it is digital economy impacts US rural agriculture. Future diversity economic base can encourage US farming product agriculture to enter rural digital agricultural service consumption sector to change US agricultural consumer individual shopping habit. What will impact to US farmers? Digital agricultural economy will bring these questions: Who will lead in broadband digital agricultural technology? How to launch e-markets and businesses and how to cooperate with US rural farmers? What will impact on US agricultural on natural resource management aspect and on how to US commodity to build amenity strategies? However, rural America's future will be shaped by policies that will encourage agricultural technology adoption, it will enhance farming worker skills, it will improve rural quality of life and it will ensure access to capital to bring US agricultural economy growth.

High level education and high birth
rate population raising technology
According to John G. (2016) indicated that " a point forecast is that GDP per capita will rise well under 1% per year in the longer run, with overall GDP growth of a little over 1 to 12%. The main drivers of slow growth are educational attainment and demographics. First, rising educational attainment will add less to productivity growth than it did historically. Second because of the aging and retirement of baby boomers, employment will rise more slowly than population which in turn, is projected to rise slowly relative to history." Thus, it seems that some economists believe education and born rate will influence future US employment method. They assume that the employment and growth ratio will rise and unemployment ratio will deadline of US will increase birth ratio and future there are many young people (students) can have chance to accept high education degree to raise whose education level to prepare to enter different kinds of high educational jobs market to work in the future.

Considering a more growth accounting perceptive on productivity growth. According to the analysis in Fernald (2015) which uses a multi-sector growth model for the projections. It indicated that although, the details differ and if bases projections on TFP data since 2004, it also implies a preferred point estimate of 1.6% per year in US over this short period. The " fundamentals" of labor-productivity growth (namely, growth in total factor productivity , TFP) have exceeded the actual realization for reasons that reflect the unwinding of dynamic of labor quality and capital depending associated with the great recession. Thus, how to raise labor quality which concerns how to raise educational level to young people in any country will be one labor economic challenge in US in the future.
Fernald (2015) also explained " the shortfall" in productivity growth relative to fundamentals reflects, at least in part, the unwinding of two dynamics associated with the great recession. He indicated that first, at the

end of the great recession business had a lot of capital relative to labor, which has attended the need of add capacity to meet demand in recent years. In contrast to the outsized growth in capital deepening from 2007 to 2010 year, we have seen capital "shallowing" since 2010 year. Second, businesses fired low-skilled workers during the recession, which raised labor quality in 2007 year to 2010 year period. As these potential workers have been rehired, the growth rate of labor quality has added less.

Thus, on the one hand, when one country encounters economic recession, many employers will like to employ high educational level and high skilled labor to raise productivity. If US encountered economic recession, but it had none (lacked) enough high educational level and high skilled knowledgeable workers to be supplied to the labor market in US. Then, US will encounter long term economic recession period, it can't shorten economic recession period. Thus, US will need to let many US young people have effort to study to prepare enough high educational level knowledge to do different kinds of professional or high knowledgeable or high technological jobs in future US labor market.

However, US high knowledgeable labor can also assist US businessmen to raise productivity growth. Due to labor quality could be grown by high educational and increasing number of knowledgeable workers. On the other hand, moreover, if there are many US young married parents who will either choose not born any babies for next generation or choose born only one to four children. Then, future US will have a small numbers of young people to be supplied to primary schools, high schools, even universities to study. It also means the university level graduate student numbers will decrease also. So, low birth rate factor can also influence the high knowledgeable labor numbers to be supplied to future US labor market. So, I recommend that US government needs to encourage many young parents choose born next generation and provides student loans to lend to poor families to support whose children can have more chance to attempt to go to high schools or universities to study in the future.

Artificial Intelligent Socio-economic development technology

US development practitioners are increasingly aware of the role that US social and political structures play in future shaping US's development paths and results. In this context, US macro social analysis needs to understand the ways in which power relations act to circumscribe the opportunities available to poor US people to improve their situation. For example, US donor organizations need to understand of relevant US social structures, such as informal institutions or other relevant US social practices in US. This provides an entry point for understanding the broader US political environment or challenges in a particular sector or process. Furthermore, any US donor organizations need to place greater emphasis on the analysis of livelihoods and economic opportunities and their relationship to reduce the unequal of gap between rich and poor US citizen in US societies . Thus, US government needs to encourage donor organization participants choose to do the reasonable donor behavioral to aim to reduce the unfair donor spending to the unneeded donor assistance beneficiaries. To let us socio-economic has more balance chance to every US poor citizen in US society. So, it can also raise average every US poor citizen feels better quality of life in US society.

How does the rise of US exports to East Asia factor influence US economy change? Export have become an increasing important source of revenue for both national and regional forms in the United States. How does the primary growth market for US exports to influence US economy growth? I recommend the developing nations in East Asia will the developing nations in East Asia will soon rival today's industrial nations as the most important US trading partner . In view of the rapid growth of US exports and their geographic shift toward developing nations.

Some economists indicates that two tends in recent US export performance are particularly notable. First, US exports have increased rapidly relative to total US output, development with important implications for the entire economy. Approximately 130,000 US firms employing over 10 million domestic workers, export their products (US Bureau of the census,1993). Second, the geographic distribution of US exports has been shifting dramatically. The industrialized nations of the organization for economic cooperation and development (OCED) account for about 57% of all US exports . However, in recent years of the start of US exports to major trading partners in East Asia. America has been rising and now accounts for 26% of the total. Thus, economic policymakers can no longer make trade related decisions without considering their effect on US exports to East Asia market.

From 1981 to 1987 year, two factors temporarily stemmed the rising tide of 1980 year. First, the US dollar began to appreciate significantly in 1981 year as the United States enacted a restrictive monetary policy and an expansion fiscal policy. From 1981 to 1985 years, the dollar rose almost 50%, thus making US exports more expensive relative to foreign products

(Hakki & Whittakersand J, Gregg Whittaker, 1985. " The US dollar recent developments, outlook, and policy options". Federal reserve bank of Kansas city, economic review, Sept./Oct., pp.3-15.) Thus, future US will need to make good international export and import trade relationship with Asia countries, e.g. China, Hong Kong, Japan etc . countries to raise GDP income.

Distance learning education technology

In the past, there fundamental forces have shaped US work labor market, includes an increase in the returns to education. General education upgrading and the large numbers of female need to work. How educational attainment, demographics and human capital will be predicted to influence US future labor market. Some economists believe education is useful to influence US future labor market. They indicate the social returns to education policies today depend on the relative prices that labor of different educational levels will command in future US labor market; current US labor market trends appear to leave a large group behind; less educated males. The rationale for social policies target specifically to this population is strengthened if predicted future outcomes in US labor market will lead this less educated male group numbers to fall down.

How to supply educational level components to influence future changes in US labor supplied? It can change in the size of the US working age population, it can change in hours worked conditional on being of working age, and it can change in the skills (effective human capital units) to US workers of different education levels, gender and age . In US , the labor is supplied by both highly educated men and women increased substantially relative to the supply of labor by the less educated. Among US males this is largely , due to an increase in educational attainment; highly educated US males did not differentially increase their supplied compared to less educated males nor did their experience differential increases in their human capital policy in US. For US females, some economists found large increases in the labor supplied by US working age women that are due to both large increases in hours worked and increase in educational attainment. So, I assume that future US will have many high educational level female labors to work in US society.

Today, human history is at the beginning of a growth industrial revolution. Developments in genetics, artificial intelligence, robotics, nanotechnology, 3D printing and biotechnology will be popular to influence US future job market change. For example, smart systems, product-homes, factories, farms or cities with help to solve problems ranging from supply chain management to climate change. The rise of the sharing economy will allow US human to monetize everything from their empty house to their car.

Due to the future patterns of consumption will change to trend high technology life enjoyment, it will cause production and employment will also change to employ high technological production labor. As entire industries adjust, most US occupations are changed to high technological manufacturing industries. When some US jobs are threatened by others grow through a change in the skill sets required to do them. a key element in understanding how the benefits and burdens of the growth.

Nowadays, the current technological changes of humans and machines , but rather an opportunity for work to truly become a channel though which human recognize full potential. As US is a high technological developed country. I assume many US employers will choose to act to be the first high technological and invention manufacturing leader to encourage which labors need to learn high technological production methods to prepare manufacture any new technological products to sell in the future US domestic or foreign both markets. Thus, it is possible that future high technological manufacturing labor numbers will be shared large go e.g. 50 to 60% future total US labor market.

How can future US labor market affect job creation and productivity growth? US future economic growth requires factor reallocation across US firms and continuous replacement of technologies. US labor market influences US economy dynamism by their impact on the supply of a key factor, skilled workers to US new and expanding firms, US growth -favoring labor market includes portable pension plans and health insurance united to the current US employers, individualized wage-setting and US public income insurance systems that encourage mobility and risk taking.

US future economic growth arises as production shifts from less to more successful firms though the reallocation of factors of production . US labor market can advance restructuring. Overly regulations tend to create a system in which a large share of economic activity occurs in US small firms without the ability to grow. US labor market should be organized to promote potential high growth US firms, especially through decentralized and individualized wage setting, portable jobs.

In the future, US will have many key importance of high growth firms. US capitalism entails a process of creative destination. New ideas continuously challenges act structures, giving rise to structural transformation as successful innovations and new products firms and industries will arise and obsolete ones will decline in US society.

Martin , J. P. (2012) studies pointed to high-growth forms (sometimes known as gazelles) as the main drivers of this process. In the US , an estimated 1% of firms creation 40% of all new jobs and 5% create almost 70% of new jobs. A review of the studies of US firms growth reveals some common findings . US high-growth firms are crucial to net job growth, generating a large share of all net jobs . This is particularly pronounced in recessions, when US high-growth firms continue to grow when other US firms deadline. US small firms are over-represented among high-growth firms, but these US firms come in all sizes . A small subgroup of large high growth firms are major job creators. Such as US high -growth firms are younger on average, US young and small high high-growth firms grows, not through mergers and acquisition and make a larger contribution to net employment growth than do US larger and older higher growth firms, high growth firms are present in all industries. Through they are slightly overrepresented in service industries in US.

Some economists predict that future US will be a flexible labor market, the marginal product of labor and the average wage in an industry should tend toward equally across US firms. Taking advantage of a legislative change to raise cost to US employers a study measured the gap between the marginal product of labor and the average wage in an industry before and after the reform. The gap increased after the legislation, which suggests that the legislation reduced allocative efficiency. Their studies have suggested that total factor productivity could increase by as much as 30% in China and India of they were to attain the US level of allocative efficiency across firms within individual industries. The result implies that plants with low total factor productivity are too large and plant with high total factor productivity are too small relative to the US benchmark of allocative efficiency.

What is allocative efficiency meaning? It occurs when the mix of product produced matches consumer preferences (where marginal benefit equals marginal cost). There products and services are the most profitable, thereby promoting economic growth other research also indicate a strong quantitative effects of strict employment protection legislation on the rate of reallocation in US industries experiments. By relaxing employment protection rules to US developed countries, such as UK, UK etc. with the strictest legislation could increases their reallocation rate by an estimated 50% in the most dynamic sectors, those that benefit most from flexibility.

The effect appears to be particularly strong on the entry-exist margin, which is arguably, especially importation for creative destruction. In future US, if manufacturing industry can have high technological production method to achieve reallocation rate to efficiency to every manufacturing industry. It can have benefit to economic growth, due manufacturing process can be move efficient to avoid cost. Also high technological manufacturing reallocation rate innovation method can influence the future US number of jobs lost in contracting or existing US firms as well as the number of jobs gained in new or expanding US firms in future a certain period divided by the average number of existing jobs. So, US high technological manufacturing reallocation rate innovation method will bring disadvantages to cause many contracting or existing firms job lost, but it will also increase the number of jobs gained in new or expanding firms in US. Thus, if future US manufacturing industry can have high technological reallocation rate innovated successfully. It will raise many new job chances in new or expanding firms in US, but it also have chance to cause many contracting or existing firms job lost or the same time.

What will expected to impact of future computerization on US labor market outcomes? Some economists estimated, about 47% of total US employment is at risk, due to that US wages and educational attainment will exhibit a strong negative relationship with an occupation's probability of computerization. They indicated that the poor performance of global labor markets across advanced economies has intensified the debate about technological unemployment among economists more recently. Indeed, over the past, decades, computers have substituted for a number of jobs, including the functions of bookkeepers, cashier and telephone operators (Bresnahan, 1999, MGI, 2013).

Although the extent of these developments remains to be seen, estimates by MGI (2013) suggests that sophisticated algorithms could substitute for approximately 140 million full time knowledge workers world wise. Hence, when technological process throughout economic history has largely been confined to the machine of manual tasks, requiring physical labor, technological progress in the twenty-first century can be expected to contribute to a wide range of cognitive tasks, which until have largely remained a human domain. Of course, may occupations being affected by these developments act still far from fully computerization, meaning that the computerization of some tasks will simply free-up time for human labor to perform other tasks. Nonetheless, the trend is clear: computers increasingly challenge human labor in a wide range cognitive tasks (Brynjolfsson and Mc Afee, 2011).

Thus, it seems that future US computerization job trend will influence some US traditional labor hand made manufacturing industry to divide some part job duty to let computerization work. However, it will not cause many US manufacturing labor to be dismissed. It can still keep some US manufacturing labors to US employers' needs in the future.

Popular science, technology , engineering
and mathematics

Science, technology , engineering and mathematics workers will drive US innovation and competitiveness by new ideas, new companies and new industries in US. However, US employers frequently concern the supply and availability of this kind of workers. Over the past 10 years, growth in this kind of jobs was three times as fast as growth to general jobs in US. This kind of US workers are also less likely to experience joblessness than other kinds of workers. In the future, science, technology, engineering and mathematics workers will play a key role to grow and raise stability of the US economy.

Bureau of labor statistics, ESA calculation 2010 and 2018 year. indicated these kinds of technology, science and engineering workers had been increasing 7.9% from 2000 to 2010 year growth as well as these kinds of workers had been increasing 17% from 2008 to 2018 year growth. Otherwise, other kinds of workers had been increasing 2.6% from 2000 to 2020 year growth as well as these kinds of workers had been increasing 9.8% from 2008 to 2018 year growth. Hence, these kinds of science, technology and engineering workers increasing rate and increasing level are more than other kinds of workers both twenty years.

The other occupations include positions, such as educators, managers, technicians, health-care professionals or social scientists. The science occupation divides four categories: computer and mathematics, engineering and surveying, physical and life science four categories. The reasons why these kinds of jobs will be trend popularly. We define these kinds of degree holders as persons whose primary or secondary undergraduate major was in a science, engineering or technological field. To using similar logic to what we used in our occupation selection, we exclude business, healthcare, and social science majors.

The US department of commerce, economics and statistics administration analysis showed that a science, technology, engineering and math. (STEM) degree is the typical path to a job related to those kinds of degrees or more than two-thirds of the 4.7 million (STEM) workers with a college degree has an undergraduate (STEM) degree. However, this does not necessarily mean that (STEM) field or their jobs . For example, only 35 per cent of college educated computer and mathematic workers have a degree in computer science or math. Thus, these past data explained that it is possible US science, technology ,engineering and mathematics employers will increase needs and why many US students choose to study these subjects in US. Thus, future US economic growth will depend on developing these kinds of science, technology ,engineering and mathematics industries.

Genetics, human intelligence,
robotics, nanotechnology, 3D
printing and biotechnology
technological industry

In US future, these kinds of jobs will be needed to increase development in genetics, human intelligence, robotics, nanotechnology, 3D printing and biotechnology. For example, smart systems homes, factories, farms grids or cities will help tackle problems ranging from supply chain management to climate change. The rise of US economy growth will allow US people to monetize everything from their empty house to their car in US. These new technological products development will change US patterns of consumption, production and employment adaption are also be changed by US corporations, US government and individuals.

Why will the technological revolution be broader socio-economic, geopolitical and demographic drivers of change to influence future US social economic and consumption pattern change? Future US most occupations will also be changed. When some traditional old jobs are threatened by redundancy and other new technological jobs will grow rapidly, existing jobs are also changed in the skill sets required to do them. The debate is between some economists foresee limitless new job opportunities and foresee massive dislocation of US jobs. In fact, the reality is highly specific to future US high technological production industry, region and high technological occupation in question as well as how US production workers can be raised themselves ability to actions the upgrade level of high technological production ability from various stakeholders to manage high technological production method change.

Overall, this is a modestly positive outlook of US high technological production employment across future most high technological production industries with jobs growth expected in several sectors. However, it is also clear that this need for more talent in certain job categories is accompanied by high skills instability across all job categories. Combined together, future US net job growth and skills instability result in most US businesses with face major recruitment challenges and talent shortages, a pattern already evident in the result and set to get worse over next five years in possible.

The question is how US businesses, government and individuals will react to these new technological job changes, due to talent shortage, mass unemployment and growing inequality challenges will encounter in future US society.

The current technological revolution does not need become a race between humans and machines , but rather an opportunity for work to truly become a channel through which US people recognize their potential. So, if US traditional low manufacturing skillful workers lack talent to learn new skills to prepare to do future new technological manufacturing jobs, such as 3 D printing, robotics, nanotechnology, biotechnological high technological products manufacturing jobs. Then, it will cause increasing of unemployment rate to some not talent US low manufacturing skillful workers. So, US government or high technological product industry employers need to consider this future unemployment challenge will be caused by high technological products manufacturing changing influences. It seems high technological development will cause these low manufacturing skillful workers unemployed rising numbers as well as high manufacturing skillful workers human capital shortage global challenges will exist.

In the future, the driver of changes to influence US demographic and socio-economic growth. They may include: changing work environments and flexible working arrangements. It means new technologies are enabling workplace innovations , such as remote working, co-working spaces and teleconferencing. Rising of the middle class in Asia markets. It means the world's economic center is shifting towards the Asia developing countries.

Some economists predict that Asia will be projected to account for 66% of the global middle class and for 59% of middle class consumption by 2030 year. In addition, climate change, natural resource will be constraints to a greener economy. It means that climate change is a major driver of innovation as organizations search for measures to help adjust to its effects. As global economic growth consumers are needed to lead to demand for natural resources and raw materials, over explanation implies higher extraction most and degradation ecosystem and these challenges will also impact US employment changes needs. All US government also needs to concern future global economic change influence.

US entrepreneurship innovation

Some economists believed the interplay between entrepreneur , innovation and economic geography growth in the United states of America have close relationship . Because future innovation is the driving force of growth in the knowledge economy to US . They assumed that if future US entrepreneurships chose to act as new firm formation, which will prefer to accept any new technological innovation to manufacture any new technological products and reallocate any new resources to apply to concentrate on manufacturing any new technological products within firms. For example, Bernard, et al. (2006) find that one third of the net increase in real U.S. manufacturing output between 1972 and 1997 is due to the net adding and dropping of products by surviving firms, a contribution that dwarfs that of net firm entry and exist. Clearly new firm formation is only one dimension of innovation and existing firms account for a large share of total research and development (R&D), in many industries such as pharmaceuticals. It seems US any industries will need spin off of existing operations and the acquisition of independent start-ups are now important dimensions of new process and product development. Thus, innovation is an important factor to influence US economic growth in the future.

However, it will have challenges to US different industries when the causal connections between entrepreneurship, innovation and growth arises. For example, US employment growth of large, middle and small sizes of entrepreneurships will be strongly positively correlated with US any new firms formation, but this doesn't necessarily imply that entrepreneurship causes growth . There may be one factor that causes US employment growth and US firms formation to co-vary, and it is hard to find instructions that affect US firm formation, but have no independent effect on US employment growth.

The interpretation of the correlation between US employment growth and US firm formation relates to old debates in US economic geography about whether US workers follow US firms, or US firms follow workers or there are mutually reinforcing feedbacks between US firms' and workers' locations decisions. So, some economists believed that, US any domestic geography economic growth will possible be caused by these factor , such as either US geography firms innovation factor or US geography high technological workers high level productivity factor or US geography firms innovation or US geography high technological workers high level productivity both combining factor. But the challenge will cause, it assumes that future many US entrepreneurships choose to innovate to manufacture whose products. Although, many high technological manufacturing workers employment chance will be raised, but it also cause many low technological manufacturing workers employment chance will be reduced as the same time. So, it is difficult to keep high technological and low technological both workers who have same fair employment chance as the same time in the future US society if US planned to achieve the future knowledgeable innovation economic society.

However, to achieve this innovation dream, the economists give opinions to US government which will need to encourage the foundations of US entrepreneurial policy and distinguish four broad actors: (a) individual agents who identify business opportunities and choose to exploit them. (b) new formed businesses which innovate using new knowledge and other resources, (C) the economy including all institutions that influence economic growth, and (d) US society is as the collection of all agents who are the ultimate beneficiaries of wealth creation. Within this organizing framework, US entrepreneurships will be easily shaped or changes the overall business climate to achieve knowledge economic society in the future.

Intangible assets ecommerce technology

What are the four big factors of intangible assets? Some economists indicated they include knowledgeable capital, human capital, social capital and entrepreneurs capital. Nowadays, globalization and increased competition will cause new types of pressure to influence US economic growth. So, US companies need have flexibility, the ability to immediately adapt to market developments, and pro-activism in creating future markets. The relative importance of physical growth: However, soft production factors, that is those related to personal knowledge are becoming more important to influence US future economic growth, which regards to human capital and knowledge as driving factor of economic growth in industries developed countries, such as US.

All these soft production factors can be grouped in what is known intangible assets. These assets can be defined as non-material factors that contribute to enterprise performance in the production of products or the provision of services, or that are expected to generate future economic benefits to the entities or individuals that control their deployment (Akerlof & Kranton , 2000).

Whether it has a close relationship between these intangible assets and US regional economic growth? Some economists had researched to have more reliable quantitative statistical information to do report in above four big factors how to effect on US regional economical growth influence. Their report indicated that returns from human and social capital are taken as homogeneous for US all regions. They also indicated these two sets of questions in their report. The first one, dealing with knowledge accumulation, addresses, among others the following issues:

(a) How does innovation and knowledge accumulation occur within firms and hoe does it impact on economic performance?

(b) What is the role of universities in regional, national and global knowledge accumulation processes?

The second set of questions addresses the key knowledge diffuses over space and how this diffusion impacts on economic performances.

In particular:

(c) To what extent knowledge diffusion is conditioned by spatial proximity?

(d) What is the impact of knowledge accumulation and diffusion on economic performance?

They indicated to answer these questions, having reliable measures of how innovation process occurs, of the actors that take part in it and the mechanisms that are in place is important. They pointed out that the systemic nature of US regional intangible asset demands indicators that grasp two kinds of capabilities: network capabilities , i.e. connectivity, both intra-and-inter-regional and organizational capabilities, as well as their dynamic in US. These indicators have been applied to the study of linkages and relationships between US firms and between US firms and US universities, highlighting the mechanisms through which those actors contribute to the processes of knowledge accumulation, generation and diffusion.

They also found results: How does knowledge accumulation occur within US firms and how does it impact on US economic performance? They included collaborations with competitors are most commonly undertaken at early stages of the development projects, those with buyers and suppliers are more likely to result in the introduction of new products and processes: At the same, it has been highlighted that, regardless of the industry, the most innovative firms, i.e. those more able to absorb knowledge are more likely to participate in collaborative networks.

Thus, US future entrepreneurships need to offer knowledge accumulation opportunities for US regional firms and research institutions to benefit and contribute to global network of US expects to achieve rapid regional economic growth. US's entrepreneurships also needs to explore organizational innovation, those US firms whose structure enables learning by doing, using and interacting by relying, among other things on parallel development teams. Semi-autonomous work teams and reduced management layers are more likely to introduce new products to the future US domestic and foreign both markets.

What is the role of US universities in US regional, national and global knowledge accumulation processes? US universities can heavily influence US regional, national and global knowledge accumulation processes. For instance, when collaborating with multi-national enterprises, they may affect simultaneously the three level. Their report had highlighted interesting results on the local impact of universities: when it has been found that top ranked departments are significantly associated with partnerships involving spatially close industry partners, it has also emerge geographical proximity, is nt the main driver of collaboration choices. These are found to depend largely on US firms' networks and universities' specific characteristics. Among other things, the cultural traditional of academic institution has been shown to influence the ability to collaborate with industry and commerce research.

Networks characteristics have emerged as crucial in determining academic knowledge transfer: the better the access to international networks, the higher the patenting activity, have the knowledge transfer to US industry. This implies that the set of tools of knowledge based economic development should include not only research and development, but also clever ways of supporting academic research network.

To conclude, if US expected research development can assist economic growth. Then, US academics universities and companies both need to co-operate to carrying on experimenting any kinds of research and prepare to assist any US product innovation more easily. So, the intangible asset, not as scientists, technology, US universities' and firms' research fund which is an conditional factor to influence US regional research and development economic growth in the future.

Ecommerce social economic development technology

Has it relationship between social influence and economic environment in future US? For example, the social factors that are positively correlated with the economic growth (i.e. the expected years of schooling and the life expectancy) and respectively, the factors that are negatively correlated with US future economic growth (i.e. the US population or risk of poverty and the unemployment rate).

The improvement of the US future economic environment will be an objective of the macroeconomic policy on short, medium and also long term. The importance of social factors upon US future economic growth, considering that the future used macroeconomic indicator, GDP per capita, is not most proper measure for the future US nation welfare. Due to GDP per capital fails to take into consideration some specific sectors of the US social economy, such as the black market.

Until recently, some economists rely on culture is as a possible determinant of economic phenomena. However, in current years, better techniques and more date made it possible to identify systematic differences in people's preferences and beliefs and to relate them to various measures of cultural environment suggest an approach to introduce cultural-based explanation that can be tested and are able to substantially understand economic phenomena.

The increased importance of social factors relies on a basic concept. Some theory is measured to economic growth which has wrong assumption. For example, the fiscal and monetary policies focused on increasing the national income, which lead consequently to economic growth. The reason most of economic opinions have been argued because whose opinions are based on a wrong hypothesis, according to which the nation welfare is based only on the level of income.

Can social factors influence US future economic growth? Human development history, global life expectancy has been experiencing these stages: from the industrialization process, the technologic progress, the medical evolution, the scientific research, these stages were also related to internal causes, specific to some developed countries, e.g. US developed country. Thus, the differences are significant and are linked both to US life expectancy level and the GDP /capita. Such as US population is less than China population too much. Although, US land area is near to China area. It seems US will encounter life expectancy level need to prepare its technological development to raise economy growth of opportunity. For example, Africa and Asia are still facing major economic and social issues. The access to a health life and medical services are still long terms objectives for countries with low life expectancy.

According to Harrison & Huntington (2000), the analysis of social factors helps understanding the human behavior with respect to consumption, savings, investment system, expectations and attitudes towards the economic circumstances, which also have a major impact on the economic growth. The evolutions of economic and social environment are needed for US future development. In order to eliminate the gap of living standard, outside resources and support US needs have good social indicators study plan to concern econometric model to rise poor people living standard in future US society between rich and poor people who are living in US. However, I believe the social factors include demographic and culture, population's structure factors which are one important social indicator to influence the distribution of the US social public income.

Barro & Sala-i-Martin (1996) defined culture is as the sum of symbols, meanings, habits values, behaviors and social artifacts which characterize a distinctive and specific human population group. For decades, economists and social thinkers debated the influence of population change on economic growth. Bloom et. al (2001) defined three alternative hypotheses: that population growth restrict, promotes or is independent of economic growth . Each hypothesis was sustained with strong arguments, and all the arguments mostly focused on population size and growth. The debates revealed other important issues, such as the age structure of the population, the way in which the population is distributed across different age groups.

The economists indicated that people's economic behavior varies at different stages of life, changes in a country's age structure can have significant effects on its economic performance. So, such as US, if it had a high proportion of children are likely to devote a high proportion of resources to their care, which would tend to depress the pace of economic growth. By contrast, if US's population falls within the working wages, the added productivity of this group can produce an increase in the economic growth. This is how the combined effect of this large working age population and health, family, labor, financial and human capital policies can create cycles of wealth creation. On the other hand, if US a large proportion consists of the elderly , the effects can be similar to those of a very young population; a large share of resources is needed by a relatively less productive segment of the population , which likewise can inhibit economic growth in future US society.

Further Bloom et al. (2001) analyzed the three main mechanisms of population's structure for determining economic growth (labor supply, savings and human capital) and their dependence of policy environment. They indicated that a growing number of adults will only be productive of there is sufficient flexibility in the labor market to allow its expansion, and macroeconomic policies that permit and encourage investment, people will only save if who have access to adequate saving mechanisms and have confidence in domestic financial markets and the demographic transition creates conditions where people will tend to invest in their health and education, offering great economic benefits, special in the modern world's increasing sophisticated economies.

It seems future US labor market will need a growing number of adults to supply , a stable domestic financial market to encourage US people have more confidence to save money in banks, a stable health and education sector industry can encourage US people have confidence to invest to do this kind of health and education service industry. Instead of population of working age factor will influence US economic growth. However, US people living culture will also influence US economic growth. For example, if it has a trend that there are many US young people like often go to shopping in relax time, because who feel shopping is their young group entertainment in the year. Then, the year US GDP growth will be possible raised, due to there are many US young people prefer to spend for shopping expenditure suddenly in the year. In conclude, in social environment of cultural and population of working age size will influence US future economic growth.

Online tourism technology

Today, US online tourism sale industry has always been one of America's great home growth industries. Today, more than 8 million Americans are employed in travel and tourism. For example, US domestic South Carolina, hospitality and tourism has been as the largest local industry, providing tens of thousands of families directly or indirectly with jobs. Predicting global tourism consumption needs work be increased. US will experience in a changing economy, tourism industry will provide job security to many Americans as well as the service -oriented nature of many travel positions and these jobs are difficult, if it is impossible to outsource.

Many people accept to buy air tickets from online sale channel. Will future America government play an important role to influence future US online air ticket tourism sale method to raise online travel consumption behavior? However, some have questioned in recent years whether US policies have harmed the ability of the tourism industry to expand. One of the most frequently discussed concerns involves American visa policy, when European Union allows tourists from 26 European countries and almost all of the America and Australia to visit without a visa. America policies require some travelers from friendly countries, such as Brazil, to travel thousands of miles from home to attend the in-person interview needed to secure a tourist visa. Although, the US White House has taken steps to reduce rise wait times in recent years, in some parts of the world, like Vietnam and Turkey, the US tourist visa application process can be a multi-month process (US Travel Association , 2014).

Will the visa waiver program impact driving increases in US impact driving increases in US tourist volumes? Some economist analysis found that when a typical country joins the visa waiver program, it sees a notable increase in the number of countries who chose to visit the US in the immediate years. That follow over the course of its first five years in the US visa wavier program, the number of tourists arriving from a participating country rises by 16.4%. Also their analysis indicated to expand the visa waiver program to some countries would result in $7.66 billion in additional tourist spending within a five year period. It would also create at least 50,000 American tourism jobs within 5 years. Purchases of products and services by visitors contribute foreign significantly to job creation and economic growth in US international travelers to US purchased more than $180 billion amount . In addition, tourism generates a trade surplus as foreign visitors spend more in the US. So, US enjoyed a $57 billion trade maintained a trade surplus in tourism every year. Some economists indicated that India is likely to contribute to the growth in foreign travel to the US . If India government can ensure which have enough numbers of foreign travelers are encouraged to choose to travel to US. In the future, India will be one important tourism cooperation partner to US. So, US ought achieve online travel sale strategy to promote to Indian to know what online travel advantages can give to them in the future, e.g. reducing time to buy air ticket, convenience, cheap e-ticket price, avoiding to lose air seat supply, more airlines choice providing. These are online e-ticket sale attractive characteristics to India online e-ticket sale market. Thus, US ought concentrate on promoting this kind of online travel sale service to India in order to raise US many airline e-tickets sale numbers.

Nowadays, China and India population growth rate are rapidly. Population growth can have several disadvantage effects on the economic expansion and performance to China and India. Does it influence cultural and sociological links between the following per capita income, rates, technological advancement , education aspects? China with 1.32 billion people and India with 1.1 billion people. Though, they each enjoy a large labor force advantage, several key economic factors have contributed to how the Chinese and Indian populations have grown and what differing effects that growth has had on their developing economies. Although, there two countries have high population growth rates, but the poor people has low standard of living.

So, when these two countries population sizes are large, but the standard of living will be low, and population will be reduced by either the preventive check (international reduction of fertility) or by the positive check (malnutrition, disease, and famine). It seems china and India governments think overpopulation challenge will cause social crime increasing, education competition, raising the job applicants number will increase, but the jobs supply number won't increase faster than job applicants number, and it will cause unemployment etc. social challenge in the future long term. So, why China and India governments will choose to control birth rates every year. However, in these developing countries, children are often as an economic asset, a tool to help increase agricultural production. Demand or need for large number of children is driven down by improvements in living standards and child survival and by the modern of economies. The movement away from agricultural production and into more modern industries in the manufacturing and services sectors need the necessity to have numbers of children to prepare to supply to China and India future manufacturing and services job markets.

However, US is a developed countries. The country main GDP growth source is high technological industry. Hence future children birth rate is not necessary to prepare to supply for manufacturing or service industries needs mainly. It seems US population birth rate is stable. So, overpopulation challenge won't be caused easily. Can population growth influence US economic growth? US is different to China and India. US's population growth rate is less. Hence, if US population growth rate can rise. It won't have overpopulation challenge. US needs have many young people number who can have high technological knowledge to prepare to have enough supply to do the future high technological different kinds of positions to build social economic growth in the future. Hence, US ought encourage young people choose to marry and to born next generation children to supply future different kinds of high technological jobs of needs to raise technological productivity and they (high technological manufacturing workers) will be the main production of factor to influence future US economic growth in the high technological manufacturing sector.

Thus, I believe the effect future US of population growth will influence US future economic growth as the same time. Also, it is the right time, US government needs to assist universities have enough afford resources, e.g. university fund, technological educational courses, lectures etc. to provide future US technological education students' needs in the future. So, it brings this question: What can determine US economic growth?

In global macro-economic influence, since 1973 year, per captia income growth in the US and other advanced countries has slowed to 2.2 % a year or almost half the 3.9% annual rate of the preceding quarter century . If the US had maintained the level of growth experiences in the 1950 year and 1960 year, real per capita income today would be about 71% greater than it actually is. In contrast, it has been estimated that eliminating the variability in US consumption since world war II would be equivalent to boosting current real consumption by only about 4.8%. If the choice is between long-term growth policies and further short-term stabilization policies long term clearly have the potential for higher benefits. Hence, what factors determine US economic growth (David M & Roy J, 1993).

According to traditional growth theory, the main determinants of long run economic growth are not influenced by economic incentives. Perhaps the reason why economists have neglected long time, to profession relied on a theory that offered litter scope for policy to influence important sources of growth.

Population growth technology

Some economists found one key is that education and anti-discrimination policies well designed labor market and large and/or progressive tax and transfer system can all reduce income inequality . In many OECD developed western countries, income inequality has increased in past decades. In some countries, top earners have captured a large share of overall income gains, when for other income has risen only a little. Some see poverty as the relevant concern with the type of growth enhancing policy reforms advocated for each OECD developed countries and economic growth might have positive or negative side effects on income inequality.

OECD (2011), it first highlights differences in some income inequality across the OECD and the factors driving them, such as cross-country differences in wage and non-wage income inequality as well as in hour worked and inactivity. OECD developed western countries can be divided into five groups to their pattern of inequality. For example, in five English-speaking countries (Australia, Canada, Ireland, New Zealand, the United Kingdom) and the Netherlands wages are rather dispersed and the share of part-time employment is high, driving inequality in labor earnings above the OECD average means-tested public cash transfer and progressive tax.

It seems income inequality will influence the developed countries, such as UK, Canada, Australia, New Zealand , US etc. economic growth. Although, technology change and globalization have played a role to influence the distribution of labor income. Some economists believe that any countries' policies will also influence income inequality. These policies factors include: technological education policies can increase different technology graduation rates from upper secondary and tertiary education and that also promote equal access to a well-designed different sector technology labor market policy can reduce inequality.

A relatively high minimum wage narrows the distribution if labor income, but if set too high, it may reduce employment of inequality reducing effect. It tends to reduce labor earning inequality by ensuring a more equal distribution of earnings. Job protection reforms that make permanent and temporary contracts more even in their provisions low income wage dispersion of earnings is rather mixed, removing product market regulations can reduce labor income inequality by boosting employment, policies the faster the immigrants and fight all forms of discrimination reduce inequality, progressive tax policy play a key role in lowering overall income across the OECD developed countries.

However, the redistributive fair income level between low level income labors and middle level income labors and high level income labors impact of developed countries, e.g. consumption taxes and real estate taxes tend to be regressive tax policy. Hence, it seems that reducing income inequality can cause the more fair income distribution between the high income level and the middle income level and the low income level labors. Then, it will let the developing countries or developed countries , such as US citizens feel that who can get real social welfare fairly, due to high technology development can boost economic growth in US society.

Cheap medical development technology

Some economists predict to make long term forecasts to reduce medical cost trends how will influence US economy. Also, they indicate short run cheap medical cost forecasts for first 1 to 5 years to reflect the particulars of specific groups, benefit packages, regional markets or cheap medical cost network providers and use their local cheap medical choice information and actuarial skills to improve accuracy and reasonability. They also oversight group review on the consistency of baseline assumptions with factor influencing future US patients' medical choice taste, medical technology trend component and historical annual percentage increase in medical costs, premiums, income and excess cost growth are found to influence future US patients consumption behavior. However, their research predicting expensive medical cost is the main factor to influence to reduce US patient numbers among of ward room sleeping environment, doctor loyalty, hospital entertainment and service facilities provision hospital cheap car parking charge service, cheap or free telephone call provision, hospital meal quality provision etc. different factors .

They suggest that how to reduce medical cost. They indicate many other factors, e.g. patient individual illness aging, physician supply, medical insurance plan benefit will affect the cost of a particular patient, group, organization or plan in a specific locality or over the short run. Also they indicated that additional analysis is still needed to determine if reliable estimates of separate medical insurance premiums, medical care payments, pharmaceuticals or other cost categories are possible to influence medical cost to be increase.

Hence, for long run US cheap medical cost will encourage or attract many patients prefer to pay medical cost to see doctors in US any hospitals or medical centers. Then, US hospitals or medical centers income will increase and US GDP income will also increase in medical service sectors in the future. It seems that long run cheap medical cost will influence US economic growth.

Artifical intelligent talent management technology

The next generation of US talent management practices and solutions will largely be driven by US economic evolution, demographic changes and technology advancements. These factors are dramatically influencing the way US people work, the way US companies are organized and the way talent is managed.

Some economists explained the key economic factors driving changes in talent management include: the knowledge economy is value to companies, talent is now a required strategic asset. Key changes in the future include a line between inside and outside talent that will result expansion of the talent management, globalization, such as European expansion is well-known top expansion prospect for global companies now include China, Russia and Eastern Europe and America and the rest of Asia. So , US government needs to know globalization can give what opportunities to US future .

Although, knowledge management or knowledge economy will bring advantages to US. However, it also bring disadvantages to this aspect, such as skill gap and structural unemployment will also influence US organizations structural unemployment and skill gap challenges issues will be caused between low skillful and high skillful workers, generational geographies changing will occur in US.

Although, baby boomer retirement has been top of mind for many years in the US, even more significant demographic changes are happening outside the US, where population growth rates and aging population will influence as local economies. Thus, the ability for organizations success will need global talent or effectively more talent from areas of abundance to scarcity is becoming a strategic issue to any US companies development , increased health and longevity mean that seniors are working longer enabling US organizations to keep experienced team members into their retirement years. But it also raise US workforce planning and generational challenges, workplace and diversity is increasing, a more diverse pool of talent affords new opportunity . Such as hiring workers raise productivity of needs, digitization of candidate or employee profiles can meet US business employment needs because employee talent data has been digitized and integrated into comprehensive talent profiles.

Techniques , such as attribute matching and recommendation technologies can be applied in talent management can be applied in talent management to find and match the most right candidate to do any positions in short time efficiently, with the increasing need of telecommunication in the US market penetration of smartphones and tablet devices a significant portion of the world's human potential will have access to rich web and application experiences from anywhere. Thus enables US organizations to source and collaborate on knowledge work with any part of the world into a global talent pool.

How can intellectual assets impact to US to change to knowledge economy?

The information age moved the basis of economic value from products to intellectual assets, information and the talent that develops them. It is now widely acknowledged that intangible assets, with largely consist of know-how, unique intelligent property, and patent right, drive move than 80% of the valuations of publishing US publicly traded companies.

In future US knowledge economy, US leading edge organizations will have efforts to use into the talent and intellectual capital of not only employees, but also clients, partners and the public at large in an effort to create an extended electronic community. For example, client support portals where clients can answer each other's questions, which will reduce costs by expensive client support calls for US companies in future US knowledge economy environment.

Also, knowledge economy will assist US companies to be partner and customer innovation efforts easily. Such as P&G daily product company's ability to source more than 50% its new product ideas to external innovation, driving industry-leading standards for new product launch rates. In future knowledge economy environment, it is clear that US organizations will be increasingly deriving value from talent that is outside the company.

In future US knowledge economy environment, the link between employees within the US organization and those outside, it is also driving talent management changes, particularly around sourcing, strategic workforce planning and employees engagement . Given these changes, executives in leading US companies will be increasingly focused on talent management issues, recognizing that talent, wherever knowledge economy will give knowledge management method to any US organizations to build more competitive advantages in the future. However, knowledge economy will also bring structural unemployment to US society. Unfortunately, today's global talent market is largely inefficient and characterized by a high degree if competition relative to the redistribution of talent.

One result of this inefficiency is long period of skill gaps structured unemployment is a particularly different scenario in which section skills are no longer required, not just within a US particular company, but within US an entire sector. Structural unemployment may be the future result of cyclical boom and bust cycles, offshoring of partial. In conclusion, whatever the causes of future knowledge economy to US , the result is an uneven distribution of talent relative to the available jobs in US. In part of structured unemployment in US will be caused from knowledge economy influence. However, knowledge economy will also influence US future businesses speed increases, bandwidth increases and growth in mobile internet access during US encounters the knowledge economy influence in society.

Distance learning educational quality technology
Schools can train skills of an individual and human capital for future businessmen need. It is not the only factor. Schools nonetheless have a special place, not only because education and skill creation are among that prime explicit objectives, but also because which the factor(s) most directly are affected by public policies. So, US is well established that the distribution personal incomes in US society is strongly related to the amount of education US people have had. Thus, any noticeable effects of the current quality of US schooling and the distribution of US people skills and income will become apparent some years in the future, those US students now in US school become a significant part of the future US labor force. Then, the question arises as: To whether are these skills correlated with US student's subsequent performance in the future US labor market and will the US economy's ability to grow ?
The quality of human resources are measured by scores is directly related to individual earnings , productivity and economic growth. A range of research results from the United States shows that earnings advantages , due to higher achievement on standardized tests are quite substantial . These studies typically find that measured achievement has a clear impact on earnings, after allowing for a differences in the quantity of schooling, age or work experience, and for other factors that might influencing US people earnings. In other words, for those leaving school at a given grad, higher quality school outcomes (represented by test scores) are closely related to subsequent earnings differences. So, it seems quality of education and individual income has direct relationship to influence US economic growth.

According, educational program that deliver those skills will bring higher individual economic benefits. Obviously, students who don't better in schools as evidenced by either examination grades or scores on standardized achievement tests tend to go further in school or university. By the same way, the net costs of improvement in school quality if reduction in rates of student repetition study. Then, it brings this question: Will education quality bring future economic return to US?

Early some economists' research found that personal and behavioral traits, such as perseverance and leadership qualities had a significant influence upon labor market success, including earnings. To answer above question, we need to know if US schools which generally have good student performance, it will influence socio-economic advantage improved performance, changes in school climate, teacher morale and commitment, school autonomy , teacher-pupil relations and disciplinary had some compensatory influence towards greater equity. In American, pupil socio-economic background and classroom climate appeared to be the most important predictors of achievement. So, it seems human development will influence US economic growth in the future. So, US development organizations seem better will then non-specialized educational organizations to design and deliver effective combinations of livelihoods and literacy.

Whether people who attain literacy actually make much use of it is subject to debate. On balance, however, literacy seems more used where economic development is better established. This supports the argument that same degree of economic and political improvement is necessary to sustain literacy. Then, US people will use their literacy skills where conditions make it useful or desirable for them to do so. However, schooling is a social process, and improvements in resources, technology and quality of student and teaching inputs should in principle to able to be enhance its overall quality. In a good number of countries, large increases in average real expenditure per student and other measures of school resources in primary and secondary schools over the last four or five decades have not remotely been matched by a comparable increase in average test scores.

What can be influenced from quality of education to US? Some economists suggest the earning can provide an important incentive mechanism, which can influence both the quality and motivation of teachers. If teachers' real average earnings had kept the same level with other professional groups over the period, the productivity impact of their earnings growth would literacy have been small. The first view point, in fact, however, in many countries teachers' earnings have increased considerably less sharply than those other groups. Teachers may feel worse off, because of their decline in status relative to other professional groups. This circumstance could well explain part perhaps an important part of the apparent lack of impact on learning outcomes of increase in real per student spending over time. One to teachers' general salary is low to compare other professional in the global society.

Then, it brings this question. Can quantity of education influence US economic growth? In this modern economic approach to investigating the determinants of educational outcomes has borrowed well established technique from other economic applications. The idea that there is a determinate relationship between (teacher number) inputs to a production process (teaching process) and the outputs(student examination results and learning performance) that subsequently bring has being been important in micro-economic analysis. If US schooling teaching possibilities are governed by certain teachers' quality between factors of production, e.g. teachers; teaching performance , teachers' supply numbers to every schools, teachers' educational qualification. So, the product function describes the maximum feasible output (student examination results and learning combinations) obtained from alternative combinations of these inputs (teachers' quality teaching performance, supply numbers, education qualification).

Production factors are powerful analytic tools, which have been applied to the analysis of most forms of economic production. Since the mid-1960 year, they have also been widely used in the economic analysis of education. Thus, in schooling educational organizations, teacher salaries will influence teachers' performance (production of factor) to be good or bad. If teachers' educational performance is good, then production of output (student learning abilities and students examination results) will be good usually. Otherwise, if their educational performance is bad, then production of output (student learning abilities and students examination results) will be bad usually .

In US society, if many poor learning performance students have been produced. Then, there will not many talent young people who can contribute to serve US society in different professional aspects. It means that they can not contribute US economic growth easily. Thus, it seems that US quality of education can impact US economic growth for long term and US universities, secondary schools and US government ought need to find methods how to raise quality of education for next generation of labor supply market.

Bio-medical technology

In the future, the perspective of bio-medical industry will be an alternative growth scenario. So, US government will need to modest improvements in key policy areas to adopt bio-medical industry needs, e.g. more favorable coverage and payment policies for medical innovation, improvements in regulatory policy to create efficiencies in research and development process and improvements in policy to incentive R & D (research and development) create efficiencies in the R & D process.

As an industry rooted in science and advanced manufacturing, the innovation bio-medical industry is uniquely position to help maintain US leadership in new technologies and scientific to continue to create high quality, high wage R&D and manufacturing jobs and enhance America's global competitiveness in the future. Today, the US bio-medical industry supports a total of 3.4 million jobs across the US economy , including over 810,000 direct jobs: contributes $189 billion in economic output and is responsible for about one in five dollars spent on domestic R&D by US businesses. Hence, innovation bio-medical industry will be one high value knowledge-based industries as a driver of US economic growth.

The reasons include: The first reason is such as , the US history, it has been the world leader in bio-medical research and development of new medicine over the past 30 years. It has been one leader of world class life science ecosystem and innovation among different countries in medicine development history. So, bio-medical industry will be a high technological innovation related activity of industry leader, as is measured by R&D investment, e.g. generation venture capital and share a total R&D employment in manufacturing industry. Next reason is that US bio-medical industry employs a total of 813, 523 worker amount. These workers lead a wide range of occupations that offer high wage, high quality employment. The total economic impacts of the industry, it is estimated (via the generally accepted methodology of input/output analysis) that supports nearly 3.4 million total jobs and generates nearly $ 789 billion in US economic output (National Science Board, 2014). Finally, reason is that US people concern life health issue, In addition to improving individual health and lengthening life ages, medical advances have contributed to substantial societal health gains, such as reducing disability and improving productivity.

According to two Universities of Chicago economists indicated the estimated economic gains from declining morality alone in the US from 1970 year to 2000 year had a value to US society of more than 3 trillion a year. Hence, US medical health need of patient numbers will increase, due to who have large needs for life health. Hence, quality of education can influence bio-medical industry development to impact US economic growth in the future.

CHAPTER THREE

How economists measure technological development between developed and developing countries

How can economists measure and think consumer behavior? It is one interesting question to assist any businessmen to apply behavioral economic psychological methods to predict how and when and why their consumers psychological

changes. Today, globalised world economics affects every aspect of human (consumers) our shopping model changes in our

daily living, such as business owner decision, investor investing behavior, employee working behavior,consumer shopping behavior.

I shall explain how economists apply behavioral economic methods to predict consumers behavioral changes in their view point.

(1) Raising consumers buying effort

Limiting free market mechanism of supply of demand

Nowadays, economists think that our society will be better served to bring consumptive benefits to our society by allowing consumers

to pursue themselves instinsts and consumptive desires to concentrate on one kind of product feature rather than by allowing different

kinds of product features to provide them to choose, e.g. in high technological fan product market, fan maufacturers ought concentrate on invent one kind of high technological fan product more than invent

different kinds of high technological fan products to let fan consumers to choose. Because lot of different features of choice to one kind of fan which will influence consumer individual spend longer time to make final shopping decision, consequently, they won't choose to buy the brand of products because the product manufacturer supplies too many of
the kind of product choices to influence their final product decisions. Thus, one kind of product choice will be the best product sale method to
adapt consumers purchase need.

Hence, it is the invisible hand comprised of consumer individual self interested shopping choices that leads to the best economic outcomes
rather than the product manufacturer's central planning(product choice planning).

The market mechanism method to judge the price of the product

Economists think any product ought follow this factor to measure or evaluate whether what the reasonable price to the product, such as:
When the quantity brought to market is just sufficient to supply the effectual demand, and no more, the market price naturally comes to be either
exactly , or as nearly as can be judged of, the same with the natural price. Thw whole quantity upon hand can be disposal of for this price, and can not be disposed for more. IN Common, commodity prices were continually towards a " central" price. In other words, the equilibrium price where supply meets demand.

Thus, the idea of a market price set the stage of the demand and supply graphs that are the heart of microeconomics. Moreover, monopolies should not be allowed to develop as they had to higher prices. For example, a shopper needs to think about how much more enjoyment they will get from buying a bit more.

Imagine it is a hot day and you see a bar serving cold drinks. The first drink will be wonderful as it helps quench a thirst and cool you down. If you have second, that will be nice but not quite as nice as the first. Have a third, and you might feel that you ought not buy to drink more. When, you might have seen the first drink as
worth the $3 price , the second one might have given you $2 of pleasure and so on. You may enjoy each drink , but the extra amount of enjoyment you get is slightly less each time. It called the " diminishing marginal utility". Hence, the principle could be applied to alomost eery field of economic

inquiry to judge whether

the product ought sell how much price is the most reasonable.

IN economics , it is frequently necessary to adopt measures, for example, measures of the general level of prices or production, which are not unique and can not be obtained from alternative similar measures by linear transformations through the origin. Such measures usually take the form of index numbers, e.g. What will be the price of the product in five years' time? Whether a raise in the price of a commodity will, given money income and the other prices, tastes, raise or lower the amount demanded ; Whether a rise in the exchange rate will raise or lower an adverse balance of payments to influence consumers choice to buy the product?

Economists also assume that government's participation can also consumer behavioral changes? In the sort of mixed economy in which we live to-day the situation, still supposed ideal, lies somewhere between these two extremes. In the old consumer behavioristic relationships continue to exercise a determining influence on certain variables.

Other variables, by legal measures and in other ways, are effectively put equal to a constant. The more the government believes in controls rather than in old mechianism , the more, if they feell all is not well, will governments tend to suppress some of the old relationships and replace them with new ones. The direct method will not always server and in such cases the government will try to alter old consumer behavioristic relationships when leaving them in their old consumptive form. The present policy as regards

saving is a mixture of these two approaches. Fiscal policy is intended to ensure among other things that government saving shall be equal to a certain among other things that government saving shall be equal to a certain constant when at the same time efforts are being made in various ways to encourage the country's people to increase their

saving. SO, their desires to consumption will be influenced to reduce. So, it seems that government's policy participation or sudden policy changing, it has indirect relationship

tp influnce consumers desires or shopping needs.

(2) Raising employees productive efficiency

Nowadays, economists recommend raising productive efficient factors , which include these three factors as below:

Each worker becomes more skilled in his or her particular contribution over time, time is saved by workers not having to swap machines and equipment as they go through each individual task and manufacturer needs to encourage how to design of machines to be better or to be improved that make it easier to do the work

for hie/her employees in factory or office or any workplace.

Also, economists think different rates of pay is particularly given he was living in an economy based more on agriculture and raw commodities than on mechanised labor. They also indicates these five factors to help any businessmen to judge how to pay their employees fairly, they may include as below:

How the employee agreeable or disagreeable the job is to pay, whether the job is easy to do or whether it requires difficult or costly training, low constant the work will be, i.e. whether it is ful-time or only intermittment, the level of trust that the employer is putting on the worker and the likelihood that the worker will be successful in the job these five factors.

So, any employer can follow above these five factors to judge whether he/she ought pay how much to the employee to let him/her to feel more fair in order to raise whose productive efficiency or improve service performance .

Division of labor and outsourcing manufacturing method can raise efficiency

For businesses, division of labor has been centralk to production systems since Ford Model-T car became the first sutomobile to be mass produced on moving assembly lines using pre-manufactured parts. More recently this has led to the growth in outsourcing, which both magnifies the specialisation of the job of making individual parts and also exploits different countries' absolute advantage in what they do. For example, Toyota says that a single car has about 30,000 parts, counting every part down to the smallest screws. When some of these parts are made at Toyato, it has a range of suppliers that make many of them.

So, real wealth is the sum of the annual produce of the land and labour to the manufacturer and it has relationship to bring real wealth to its whole country income rather than the monetary wealth.

Surplus value of labor is a waste production raising cost to manufacturers

Value raising from the effort that workers put in to produce the products, the amount of labour used to make products determined their prices over the long run, Econoimists focus on value. IN their view the amount of labour determined the value of the product produced. For example,a machine take takes five hours to make something has twice the value as one that takes ten hours. They distinguished between use value , which was immutable and the exchange value that the owners of the product could get by selling it. They could workers did not retain all the value of their efforts by a long run. A machine worker would be paid a wage to produce a box-worth of tools that would sell for a much higher price than his daily wage. Think of a supermarket checkout worker, a solicitor's assistant or a bank teller , they ought need to be pay higher salary to compare the staffs who are not assistant to their organizations, because their job duties are more complex and difficult to compare others in their organizations, instead of their managers.

Welfare economic theory, Kaldor-Hicks efficiency theory and Pareto improvement theory relationship
Economists explain that welfare economic theory, Kaldor-Hicks efficiency theory and Pareto improvement theory which have close relationship. Welfare economic theory means general social equilibrium. They indicate the demand of commodities must same to the supply of commodities when the commodity market stays in the stable level. So , when production is increasing, the manufacturers also need to keep their products supply and demand keep to the stable level and consumers' demand of their products must also need to increase as well as bank interest rate must be fallen in order to raise consumption desires. Also, the low bank interest can also excite investors' investment desires.

Such as to explain why the developing countries cities technological competitive investment factor can impact US and UK economic growth. Welfare economic theory means general social equilibrium. The economists indicate the demand of commodities must same to the supply of commodities when the commodity market stays in the stable level. Due to the future developing countries investors will invest much on developing new technological products research. So, these countries' economy growth will be possible faster than US and UK both developed countries because it is possible that the technological product investor number will decrease. So, US and UK government will need to attract and encourage many foreign developing countries' investor prefer to invest their technological products research to assist US and UK both developed countries to develop its economy. As Welfare economic theory means general social equilibrium. If US and UK government can give benefits to these developing countries' technological product investors, then these developing countries' technological product investors will prefer to invest to developed country US urban cities, such as small cities or farming location to build factories to manufacture their new technological products. Because , London, New York, Washington etc. large cities' technological investment and location where land supply is limited and offices and factories rents are also expensive. Otherwise, UK and US, farming location or small cities location where have much land supply to build large offices and facilities and the rent is much cheaper than US and UK main cities, such as London, New York, Washington etc. Even, if UK and US government and banks can lend low interest rate of bank loans or government loans to encourage foreign investors to do technological product research in its country. So, UK and US bank low bank interest rate can encourage US consumers prefer to buy any future new technological products, due to low bank interest rate can not attract who choose to save more money in US and UK any banks as well as US and UK bank low bank interest rate can also encourage foreign developing countries' high technological product investors to choose to invest in UK and UK , due to who can pay low bank or government loan interest to compare themselves countries or other countries.

As Kaldor-Hicks efficiency theory indicates the policy is suitable to be implemented if the policy beneficiaries' welfares can compensate to the policy benefactress. So, US and UK bank and government low loan interest rate policy is valuable to implement if benefactress, such as US and UK government which can raise economic growth as well as the US benefactress, such as the high technological product businessmen can sell many different kinds of high technological products from foreign developing countries' investors' assistance in US and UK both developed countries.

As Pareto improvement theory indicates that it can not carry on continue, due to the country has shortage of natural resources after the country improved its welfares to reach the maximum standard. The result will be caused, such as the country will reduce this group of citizen's welfares, if the country decides to continue to improve another group of citizen's welfares. So, future US and UK will have shortage of natural resources to supply to technological product manufacturers. So, the number of technological product production will be fallen , due to the shortage of natural resources are supplied to US and UK high technological product manufacturers. The only solvable choice is encouraging foreign high technological product investors to attract them to supply themselves high technological products any resources to manufacture high technological products and provide to US and UK high technological product sellers to help them to sell in US and UK domestic market or export to overseas market to earn income. Then, US and UK will earn raise GDP income from high technological product sale industry in the future.

Natural rate of unemployment factor

Economists believe why natural rate of unemployment will occur in any countries. They indicate that during economy condition keeps at the balance situation (condition), anyway any country government adopts any policy. However, the natural rate of unemployment won't reach to zero level as well as there are some people will still unemployed or it is possible that some people will be alternative employment between any time. Hence, it means that although US and UK is an developed country. It can not guarantee there are not any people unemployed , so the natural rate of unemployment will not reach to zero level.

So, it seems that US and UK current economy condition had kept at the balance situation (condition). Thus, even future science, technology , engineering and mathematics workers will drive US innovation and competitiveness by new ideas, new companies and new industries in US. However, US and UK employers frequently concern the supply and availability of this kind of workers. Over the past 10 years, growth in this kind of jobs was three times as fast as growth to general jobs in US. In the future the Natural rate of unemployment of US and UK science, technology , engineering and mathematics worker number must not reach zero level.

Anyway, in the future, the perspective of bio-medical industry will be an alternative growth scenario. So, US and UK government will need to modest improvements in key policy areas to adopt bio-medical industry needs, e.g. more favorable coverage and payment policies for medical innovation, improvements in regulatory policy to create efficiencies in research and development process and improvements in policy to incentive R & D (research and development) create efficiencies in the R & D process.

As an industry rooted in science and advanced manufacturing, the innovation bio-medical industry is uniquely position to help maintain US leadership in new technologies and scientific to continue to create high quality, high wage R&D and manufacturing jobs and enhance America's global competitiveness in the future. In the future the Natural rate of unemployment of US and UK of bio-medical industry worker number must not reach zero level.

Thus, if future US had enough job number which could supply to these high technological and bio medical industries of labors to do. However, the natural rate of unemployment must not reach at zero level and it does not represent its economy is poor because one developed country , such as US had been experiencing economy condition keeps at the balance situation(condition), so it's natural rate of unemployment must not reach zero level . Otherwise, the developing countries' economy condition does not keep at the balance situation (condition), so their natural rate of unemployment have more possible to reach to close zero level.

Industrial production and pollution economic factor

Economists indicate that it has close relationship between industrial production and pollution. When traditional products are manufactured, the air and water pollution will be caused in the economic activity. So, human needs to protect environment to keep health, the demand of investment of money reduces pollution and labor number will increase to achieve to reduce air and water pollution in natural environment. Thus, product manufacturers need to analyze which economic stage(s) will encounter shortage and find methods to solve challenge.

This Industrial production and pollution economy theory can explain why US and UK needs to plan achieve digital economy to shift future America rural economy. Some economists indicate that there are five trends reshape to impact rural America' and England future economy. They include that digital economy will shift future America rural economy. US and UK quality of life will change a lot, the US and UK rural economy will stay uneven, US commodities will compete in global markets and will give less benefit to US and UK rural economy and US and UK new products will revolutionize US and UK agriculture economy.

US and UK needs to plan digital economy to future agricultural production of the reasons include: The first aspect impacts to US agriculture economy, the US future rural economy stays uneven. Growth will concentrate in 4 out of 10 rural places and they have scenery, a retail hub, or one next to a city in US and UK. The impact on rural America includes some rural places will try to manage growth , but many places on a quest for new economic engines. Thus, it will bring these questions to US and UK rural economy impact, such as : Who will be US and UK businessmen clients? The struggling farmers? The struggling farm-dependent country? The booming mountain area? The rural area transforming into city? The second aspect impacts to US and UK agriculture economy, due to US and UK commodities will compete in global markets if it will bring a smaller benefit in US and UK rural economy. Then, US and UK farm scale will cut costs and competition fewer US and UK farms and places will depend on farm income. The third aspect impact to US and UK agriculture economy, new products will revolutionize agriculture. Major, shift from commodities to products, spurred by biotech, means two agriculture in the future and two rural America . Hence, US and UK future agriculture determination the rural economy will be declined. A future new US and UK agriculture supply chain integrator will be caused from the traditional farming supply chain procedure the change to outsourced contractor farming supply chain procedure, such as: In beginning, from farmer will outsource supply chain contract to processor, then contract to distributor and contract to food retailer final step. The fourth aspect impacts to US and UK agriculture economy, what will be two agricultures impact to the US and UK future economy? The first US and UK agriculture impact will be US and UK commodity agriculture. It focuses on production capabilities, farming foods production will be thin margins maintained with technology and big sale. The second US and UK agriculture impact will be product agriculture. It focuses on consumer needs, farming foods production margins will be protected by capturing value and building business relationship. In the fifth aspect impacts to US and UK agricultural economy, it is digital economy impacts US and UK rural agriculture. Future diversity economic base can encourage US and UK farming product agriculture to enter rural digital agricultural service consumption sector to change US and UK agricultural consumer individual shopping habit.

Thus, future US and UK agricultural production needs to concern how to reduce air and water pollution to influence natural environment clean quality as well as it needs to find what economic stage of agricultural production will be innovated in the future.

Reference

Akerlof, G.A. & Kranton, R.E. (2000): Economics & Identity, Quarterly Journal Of Economics, 105(3) , 715-753.

Barro, R., Sala-i-Martin, X., (1996). The classical approach to convergence analysis. The economic journal, vol. 106, no. 437.

Bernard, A., Redding, S & Schott, P. (2006). " Multi- product firms and product switching, " NBER working paper, 12293.

Bloom, D., Canning, D., Sevilla, J., 2001. Economic growth and demographic transition. National Bureau Of Economic Research.

Brynjolfsson and Mc Afee, (2011). Race against the machine: how the digital revolution is accelerating innovation and driving productivity , US.

Bresnahan, T.T. (1999). computerisation and wage dispersion: an analytical reinterpretation. The economic journal, vol. 109, no. 456 pp. 390-415.

David. M Gould & Roy. J. Ruffin. What determines long run economy growth? Economic review, second quarter, 1993.

ESA (2010 & 2018 year) calculations using current population survey public use micro date and estimates from the employment projections program of those Bureau of labor statistics.

Fernald, John (2015). " Productivity snd potential output before during, and after the great recession" In Jonathan A Parker & Michael

Hakkio, Craig S. 1992. " Is purchasing power parity a useful guide to the dollar?" Federal reserve bank of Kansas city, Economic review, third quarter, pp. 37-51.

Harrison, Huntington, S . (2000). Culture matters: how values shape human progess. Basic books.

John, G. 2016 " Reassessing longer run US growth. How low? " Federal reserve bank of San Francisco working paper 2016-18. http://www.frbsf.org/economic-research/publications/working-papers/wp2016-18/pdf.

Martin, J.P. & S. Scarpetta. " Setting it right. employment protection, labor reallocation and productivity." De Economist , 60: 2 (2012): 89-116. Online at: http://ideas. repec.org/p/iza/izapps/pp.27 html (3).

MGI (2013). Disruptive technologies: Advances that will transform life business and the global economy. Tech. Rep: Mckinsey Global Institute.

National Science Board, 2014 , Science and Engineering Indicators, USP TO Patent applications and grants by industry.

OECD (2011), Divided We Stand: Why Inequality Keeps Rising, OECD publishing.

US Travel Association " US Travel Employment Reaches An Time high" (press release) , Nov. 7, 2014 Accessed Dec. 8 , 2014. available here : http://www.ustravel.org/news/press-releases/travel-industry-employment-reaches-all-time-high.

US Bureau of the census, 1993. Statistical abstract of the United States, 113 the ed. Washington.

CHAPTER FOUR

Measurement consumers behaviors and thinking consumers shopping psychological changes as well as raising employees efficiency

How can economists measure and think consumer behavior? It is one interesting question to assist any businessmen to apply behavioral economic psychological methods to predict how and when and why their consumers psychological
changes. Today, globalised world economics affects every aspect of human (consumers) our shopping model changes in our
daily living, such as business owner decision, investor investing behavior, employee working behavior,consumer shopping behavior.
I shall explain how economists apply behavioral economic methods to predict consumers behavioral changes in their view point.

(1) Raising consumers buying effort

    Limiting free market mechanism of supply of demand

    Nowadays, economists think that our society will be better served to bring consumptive benefits to our society by allowing consumers
to pursue themselves instinsts and consumptive desires to concentrate on one kind of product feature rather than by allowing different

kinds of product features to provide them to choose, e.g. in high technological fan product market, fan maufacturers ought concentrate on invent one kind of high technological fan product more than invent different kinds of high technological fan products to let fan consumers to choose. Because lot of different features of choice to one kind of fan which will influence consumer individual spend longer time to make final shopping decision, consequently, they won't choose to buy the brand of products because the product manufacturer supplies too many of the kind of product choices to influence their final product decisions. Thus, one kind of product choice will be the best product sale method to adapt consumers purchase need.

Hence, it is the invisible hand comprised of consumer individual self interested shopping choices that leads to the best economic outcomes rather than the product manufacturer's central planning(product choice planning).

The market mechanism method to judge the price of the product

Economists think any product ought follow this factor to measure or evaluate whether what the reasonable price to the product, such as: When the quantity brought to market is just sufficient to supply the effectual demand, and no more, the market price naturally comes to be either exactly , or as nearly as can be judged of, the same with the natural price. Thw whole quantity upon hand can be disposal of for this price, and can not be disposed for more. IN Common, commodity prices were continually towards a " central" price. In other words, the equilibrium price where supply meets demand.

Thus, the idea of a market price set the stage of the demand and supply graphs that are the heart of microeconomics. Moreover, monopolies should not be allowed to develop as they had to higher prices. For example, a shopper needs to think about how much more enjoyment they will get from buying a bit more.

Imagine it is a hot day and you see a bar serving cold drinks. The first drink will be wonderful as it helps quench a thirst and cool you down. If you have second, that will be nice but not quite as nice as the first. Have a third, and you might feel that you ought not buy to drink more. When, you might have seen the first drink as worth the $3 price , the second one might have given you $2 of pleasure and so on. You may enjoy each drink , but the extra amount of enjoyment you get is slightly less each time. It called the " diminishing marginal utility". Hence, the principle could be applied to alomost eery field of economic inquiry to judge whether

the product ought sell how much price is the most reasonable.

IN economics , it is frequently necessary to adopt measures, for example, measures of the general level of prices or production, which are not unique and can not be obtained from alternative similar measures by linear transformations through the origin. Such measures usually take the form of index numbers, e.g. What will be the price of the product in five years' time? Whether a raise in the price of a commodity will, given money income and the other prices, tastes, raise or lower the amount demanded ; Whether a rise in the exchange rate will raise or lower an adverse balance of payments to influence consumers choice to buy the product?

Economists also assume that government's participation can also consumer behavioral changes? In the sort of mixed economy in which we live to-day the situation, still supposed ideal, lies somewhere between these two extremes. In the old consumer behavioristic relationships continue to exercise a determining influence on certain variables.

Other variables, by legal measures and in other ways, are effectively put equal to a constant. The more the government believes in controls rather than in old mechianism , the more, if they feell all is not well, will governments tend to suppress some of the old relationships and replace them with new ones. The direct method will not always server and in such cases the government will try to alter old consumer behavioristic relationships when leaving them in their old consumptive form. The present policy as regards

saving is a mixture of these two approaches. Fiscal policy is intended to ensure among other things that government saving shall be equal to a certain among other things that government saving shall be equal to a certain constant when at the same time efforts are being made in various ways to encourage the country's people to increase their

saving. SO, their desires to consumption will be influenced to reduce. So, it seems that government's policy participation or sudden policy changing, it has indirect relationship

tp influnce consumers desires or shopping needs.

(2) Raising employees productive efficiency

Nowadays, economists recommend raising productive efficient factors , which include these three factors as below:

Each worker becomes more skilled in his or her particular contribution over time, time is saved by workers not having to swap machines and equipment as they go through each individual task and manufacturer needs to encourage how to design of machines to be better or to be improved that make it easier to do the work
for hie/her employees in factory or office or any workplace.

Also, economists think different rates of pay is particularly given he was living in an economy based more on agriculture and raw commodities than on mechanised labor. They also indicates these five factors to help any businessmen to judge how to pay their employees fairly, they may include as below:

How the employee agreeable or disagreeable the job is to pay, whether the job is easy to do or whether it requires difficult or costly training, low constant the work will be, i.e. whether it is ful-time or only intermittment, the level of trust that the employer is putting on the worker and the likelihood that the worker will be successful in the job these five factors.

So, any employer can follow above these five factors to judge whether he/she ought pay how much to the employee to let him/her to feel more fair in order to raise whose productive efficiency or improve service performance .

Division of labor and outsourcing manufacturing method can raise efficiency

For businesses, division of labor has been centralk to production systems since Ford Model-T car became the first sutomobile to be mass produced on moving assembly lines using pre-manufactured parts. More recently this has led to the growth in outsourcing, which both magnifies the specialisation of the job of making individual parts and also exploits different countries' absolute advantage in what they do. For example, Toyota says that a single car has about 30,000 parts, counting every part down to the smallest screws. When some of these parts are made at Toyato, it has a range of suppliers that make many of them.
So, real wealth is the sum of the annual produce of the land and labour to the manufacturer and it has relationship to bring real wealth to its whole country income rather than the monetary wealth.

Surplus value of labor is a waste production raising cost to manufacturers

Value raising from the effort that workers put in to produce the products, the amount of labour used to make products determined their prices over the long run, Econoimists focus on value. IN their view the amount of labour determined the value of the product produced. For example,a machine take takes five hours to make something has twice the value as one that takes ten hours. They distinguished between use value , which was immutable and the exchange value that the owners of the product could get by selling it. They could workers did not retain all the value of their efforts by a long run. A machine worker would be paid a wage to produce a box-worth of tools that would sell for a much higher price than his daily wage. Think of a supermarket checkout worker, a solicitor's assistant or a bank teller , they ought need to be pay higher salary to compare the staffs who are not assistant to their organizations, because their job duties are more complex and difficult to compare others in their organizations, instead of their managers.